Restoring Communities Resettled Construction in Asia

The rapid economic expansion and population growth of developing countries in Asia has led to increasing demands for water and energy. To meet these demands, large dam development projects have been completed, which has inevitably caused involuntary resettlement. In order to support these projects, dam developers must find appropriate ways to ensure adequate livelihood reconstruction for resettled individuals. Resettlement causes both short-term and long-term effects (both positive and negative) for the relocated populations, meaning that in order to evaluate the larger impact of such projects long-term post-project evaluations must be carried out. However, post-project evaluations by international donors have typically been conducted within a few years after completion; the long-term impact of such projects is seldom evaluated.

This book aims to fill this gap. A study team composed of researchers from Indonesia, Japan, Lao PDR, Sri Lanka, and Turkey has conducted ten case studies focusing on resettled individuals' satisfaction, opportunities offered, and income generation. The volume provides an overview of the ten case studies, which were carried out across five countries. It also discusses how a compensation programme should be designed and what sort of options should be presented to resettled individuals for their maximum benefit.

This book was originally published as a special issue of the *International Journal of Water Resources Development*.

Mikiyasu Nakayaka serves as a professor at the University of Tokyo, Japan. He specializes in management of environment and natural resources, particularly water resources.

Ryo Fujikura serves as a professor at Hosei University, Tokyo, Japan. He specializes in environmental systems, environmental policy formulation and international environmental cooperation.

Restoring Communities Resettled After Dam Construction in Asia

Edited by
Mikiyasu Nakayama and Ryo Fujikura

Routledge
Taylor & Francis Group

LONDON AND NEW YORK

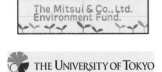

The Mitsui & Co., Ltd.
Environment Fund.

THE UNIVERSITY OF TOKYO

科研費
KAKENHI

HOSEI

First published 2014
by Routledge

Published 2014 by Routledge
2 Park Square, Milton Park, Abingdon, Oxfordshire OX14 4RN

and by Routledge
711 Third Avenue, New York, NY 10017

Routledge is an imprint of the Taylor and Francis Group, an informa business

First issued in paperback 2015

British Library Cataloguing in Publication Data
A catalogue record for this book is available from the British Library

ISBN 978-0-415-71910-0 (hbk)
ISBN 978-1-138-95388-8 (pbk)

Typeset in Times New Roman
by Taylor & Francis Books

Publisher's Note
The publisher accepts responsibility for any inconsistencies that may have arisen during the conversion of this book from journal articles to book chapters, namely the possible inclusion of journal terminology.

Disclaimer
Every effort has been made to contact copyright holders for their permission to reprint material in this book. The publishers would be grateful to hear from any copyright holder who is not here acknowledged and will undertake to rectify any errors or omissions in future editions of this book.

Contents

CONTENTS

Citation Information

The chapters in this book were originally published in the *International Journal of Water Resources Development*, volume 29, issue 1 (March 2013). When citing this material, please use the original page numbering for each article, as follows:

Chapter 1
The long-term impacts of resettlement programmes resulting from dam construction projects in Indonesia, Japan, Laos, Sri Lanka and Turkey: a comparison of land-for-land and cash compensation schemes
Ryo Fujikura and Mikiyasu Nakayama
International Journal of Water Resources Development, volume 29, issue 1 (March 2013) pp. 4-13

Chapter 2
The resettlement programme of the Wonorejo Dam project in Tulungagung, Indonesia: the perceptions of former residents
Dian Sisinggih, Sri Wahyuni and Pitojo Tri Juwono
International Journal of Water Resources Development, volume 29, issue 1 (March 2013) pp. 14-24

Chapter 3
Livelihood status of resettlers affected by the Saguling Dam project, 25 years after inundation
Sunardi, Budhi Gunawan, Jagath Manatunge and Fifi Dwi Pratiwi
International Journal of Water Resources Development, volume 29, issue 1 (March 2013) pp. 25-34

Chapter 4
Resettlement and development: a survey of two of Indonesia's Koto Panjang resettlement villages
Syafruddin Karimi and Werry Darta Taifur
International Journal of Water Resources Development, volume 29, issue 1 (March 2013) pp. 35-49

Chapter 5
A long-term evaluation of families affected by the Bili-Bili Dam development resettlement project in South Sulawesi, Indonesia

Hidemi Yoshida, Rampisela Dorotea Agnes, Mochtar Solle and Muh. Jayadi
International Journal of Water Resources Development, volume 29, issue 1 (March 2013) pp. 50-58

Chapter 6
The livelihood reconstruction of resettlers from the Nam Ngum 1 hydropower project in Laos
Bounsouk Souksavath and Miko Maekawa
International Journal of Water Resources Development, volume 29, issue 1 (March 2013) pp. 59-70

Chapter 7
Reconstruction of the livelihood of resettlers from the Nam Theun 2 hydropower project in Laos
Bounsouk Souksavath and Mikiyasu Nakayama
International Journal of Water Resources Development, volume 29, issue 1 (March 2013) pp. 71-86

Chapter 8
Long-term perceptions of project-affected persons: a case study of the Kotmale Dam in Sri Lanka
Jagath Manatunge and Naruhiko Takesada
International Journal of Water Resources Development, volume 29, issue 1 (March 2013) pp. 87-100

Chapter 9
Atatürk Dam resettlement process: increased disparity resulting from insufficient financial compensation
Erhan Akça, Ryo Fujikura and Çiğdem Sabbağ
International Journal of Water Resources Development, volume 29, issue 1 (March 2013) pp. 101-108

Chapter 10
The long-term implications of compensation schemes for community rehabilitation: the Kusaki and Sameura dam projects in Japan
Kyoko Matsumoto, Yu Mizuno and Erika Onagi
International Journal of Water Resources Development, volume 29, issue 1 (March 2013) pp. 109-119

Please direct any queries you may have about the citations to clsuk.permissions@cengage.com

The long-term impacts of resettlement programmes resulting from dam construction projects in Indonesia, Japan, Laos, Sri Lanka and Turkey: a comparison of land-for-land and cash compensation schemes

Ryo Fujikura[a] and Mikiyasu Nakayama[b]

[a]Faculty of Humanity and Environment, Hosei University, Tokyo, Japan; [b]Department of International Studies, Graduate School of Frontier Science, University of Tokyo, Japan

Post-project household surveys were conducted regarding 10 resettlement programmes resulting from dam construction projects in Indonesia, Japan, Laos, Sri Lanka and Turkey. In all cases the resettlement was completed at least 20 years ago, except for one case in Laos. Six of the programmes adopted a cash compensation scheme and the other four were based on a land-for-land compensation scheme. While the World Bank and the Organisation for Economic Co-operation and Development prefer land-for-land compensation, there was no significant difference observed concerning the effectiveness of the two compensation schemes. Cash compensation demonstrated a small advantage for farmers who wanted to change their occupation; for those who hope to move into an urban area to secure a better livelihood, cash compensation could be a better option.

Introduction

As the population and economy of developing countries in Asia have been rapidly expanding, demands for water and energy have been increasing also. The development of nuclear power has become more difficult since the accident at the Fukushima nuclear power station in Japan in March 2011. Hydropower production across the world is estimated to have a potential of 15.9 trillion kWh; only 2.6 trillion kWh (16.3%) has currently been developed (New Energy Foundation, 2010). Therefore, hydropower generation through the construction of large dams may be a more feasible renewable energy source that can meet the power demands of various Asian cities; many of these cities already face a power shortage. Laos and Nepal regard themselves as strategically positioned to become the "batteries of Asia" through their hydropower development. There is great potential for these countries to enhance their economy by selling more electricity to neighbouring countries (Laos to Thailand and Nepal to India). Many developing countries, including these nations, are seeking funds from foreign donors to develop large dams for hydropower generation and other purposes.

Large dam development projects inevitably cause involuntary resettlement. The number of resettled individuals in the world is estimated to be between 40 and 80 million. Many of these resettlement projects have failed to adequately reconstruct the livelihood of the resettled individuals and many have become impoverished (World Commission on Dams, 2000). As a result, many dam opponents have established global anti-dam networks making use of the advancement of information technologies. Dam development issues therefore have become "internationalized". That is, both international financial institutions (e.g. the World Bank) and developed countries have become more cautious in funding dam development projects because of the potential social conflict. In the meantime, China has been actively investing in large dam projects overseas. Chinese banks and companies have financed or are considering some 300 dam projects in 66 countries (International Rivers, 2012).

Dam development is almost inevitable in developing countries. In order to support these projects, dam developers must find appropriate ways to ensure adequate livelihood reconstruction for resettled individuals. Resettlement causes both short-term and long-term effects (both positive and negative) for resettled individuals; for example, employment for the second generation of resettled individuals (a long-term impact) cannot be observed within just a few years after resettlement. In order to evaluate the larger impact of such projects, long-term post-project evaluations must be carried out. However, post-project evaluations by international donors have typically been conducted within a few years after completion; the long-term impact of such projects is seldom evaluated.

A study team was organized from researchers from Japan, Indonesia and Sri Lanka in 2006 to function as a precursor to the studies in this special issue. From 2006 to 2009, the study conducted long-term post-project evaluations of six large dam projects. These projects included the Ikawa Dam, three Jintsu Dams (Jintsu 1 to Jintsu 3) and the Miyagase Dam in Japan; the Koto Pangjang, Bili-Bili and Saguling Dams in Indonesia; and the Kotmale Dam in Sri Lanka. The findings from these nine case studies were developed into a set of suggestions for design and implications for future dam resettlement programmes. Recommendations included appropriate institutional arrangement, secured implementation of the resettlement programme, and sufficient consideration of emotional factors affecting resettled individuals (Fujikura, Nakayama, & Takesada, 2009).

The team conducted 10 additional case studies that focused on resettled individuals' satisfaction, opportunities offered, and income generation. The projects studied include the Wonorejo, Saguling, Koto Pangjang and Bili-Bili Dams in Indonesia; Nam Ngum 1 and Nam Theun 2 in Laos; the Kotmale Dam in Sri Lanka; the Atatürk Dam in Turkey; and the Kusaki and Sameura Dams in Japan. Except in the case of the newly constructed Nam Theun 2, the resettled individuals had left their original home more than 10 years ago at the time of the survey. The researchers attempted to locate first-generation residents of the resettlement areas for interview, if any were still present. The interviews were conducted from 2010 to 2012 and except for the Japanese case, there were approximately 100 interviewees per project.

Throughout the study, it was recognized that the level of compensation and amount of choice given to the resettled individuals was crucial to their satisfaction. The results suggest that cash compensation is sometimes more advantageous than land-for-land compensation; nevertheless, the World Bank has maintained its preference for land-for-land solutions (World Bank, 2004, p. 61). The World Bank's operational policy specifies that "preference should be given to land-based resettlement strategies for displaced persons whose livelihoods are land-based" (World Bank, 2004). The World Bank emphasizes the high risk of cash compensation, which may be subject to loss or delay, or be used for purposes other than life rehabilitation, including consumption, ceremonial expenses, or repayment of loans (World Bank, 2004, p. 67). In addition, the Organisation for Economic Co-operation and Development has discouraged cash compensation, stating that it "typically leads to impoverishment" (Development Assistance Committee, 1992, p. 12). However, a land-for-land compensation scheme may effectively limit residents to rural farming, thus depriving them of the opportunity to choose their future. This paper provides an overview of the 10 case studies, which were carried out in 5 countries; it also discusses how a compensation programme should be designed and what sort of options should be presented to resettled individuals for their maximum benefit.

Overview of case studies

Wonorejo Dam, Indonesia

The Wonorejo Multipurpose Dam was constructed in East Java Province, Indonesia. The Asian Development Bank (ADB) and later the Overseas Economic Cooperation Fund (OECF) of Japan were the financiers of this project. Construction was completed in 2001 and the resettlement compensation was monetary. The resettled individuals took three different routes: 475 families joined the nation's transmigration programme and moved onto islands other than Java or Bali (i.e. Papua, Kalimantan, Sumatra and Sulawesi); 356 moved to surrounding villages downstream from the dam project; and 165 moved upstream of the dam project.

Sisinggih, Wahyuni, and Juwono (2013) conducted a survey of 88 resettled individuals relocated either upstream or downstream of the dam project. The respondents included both first-generation individuals who had decided to relocate and their second generation. The government had promised relocated families more land in the resettlement area as well as cash compensation. However, many of the resettled individuals (521) chose to stay near the reservoir due to strong emotional ties to the region. This same group is generally satisfied with their livelihood after resettlement. They used their cash compensation to purchase more land, to build better homes, to acquire better infrastructure and to provide higher education opportunities for their children. However, the first generation still faces difficulties in adapting to the new living environment and in achieving income stability. Those whose land was only partially submerged were unable to access full cash compensation and also faced difficulty in reconstructing their livelihood. According to the information provided by the respondents, resettled individuals who joined the transmigration programme generally succeeded in livelihood rehabilitation.

Saguling Dam, Indonesia

The Saguling Dam was constructed in West Java Province, Indonesia, and was completed in 1987. The financier was the World Bank. This project displaced 3038 families from the inundated area and affected 7626 families altogether. The latter had land or sources of income in the inundated area. The government had offered several options for these families: (1) transmigration to islands outside of Java; (2) migration within West Java; and (3) migration to a location of their choice. Only 3.9% of the resettled individuals followed one of the first two schemes; the remainder of the resettled individuals chose to relocate near the reservoir (Suwartapradja, Arifin, Kanum, Ansor, & Djumari, 1985). As a result, population density around the reservoir increased. For these resettlers, another option – livelihood rehabilitation by aquaculture in the reservoir – was provided by the World Bank. Cage aquaculture and capture fisheries were initiated to provide work for the resettled individuals. However, the level of cash compensation offered to the majority of the resettled individuals proved insufficient to purchase the necessary fish cages. As a result, the practice of aquaculture in the reservoir is conducted mostly by entrepreneurs from the outside rather than by resettled individuals (Manatunge, Takesada, Miyata, & Nakayama, 2009).

Sunardi, Manatunge, and Pratiwi (2013) conducted a survey of 147 families who had resettled from two submerged villages. The number of self-employed farmers decreased drastically after resettlement and the number of share-croppers and unemployed individuals increased. The net and cage aquaculture industry generally failed within a few years due to fish disease, poor water quality, and the 1998 national economic crisis in Indonesia. Most of the resettled individuals in both villages were not able to reconstruct the facilities required for aquaculture because of their limited access to capital and assistance. The majority of the resettled individuals, however, are satisfied with their new living environment and educational opportunities despite the fact that they are financially worse off than before. Unemployment is the most common negative impact of resettlement.

Koto Panjang Dam, Indonesia

The Koto Panjang Dam was constructed in the central part of Sumatra, Indonesia. This project was funded by the OECF and was initiated to provide electricity to the provinces of Riau and West Sumatra. Construction of this dam resulted in the inundation of 10 native villages and the displacement of approximately 5000 households; 8 of these villages were in Riau Province and 2 in West Sumatra Province. The resettlement process took 10 years (from 1991 to 2000) to complete and treated each village in a unique way in the sense that they had control over their final destination. Compensation was in terms of land and money; every family received a certain amount of land (e.g. 0.5 ha for each resident of Koto Mesjid). Cash compensation was paid according to the submerged assets of the residents.

From the 10 ten resettled villages, Karimi and Taifur (2013) selected Koto Mesjid in Riau Province and Tanjung Balik in West Sumatra Province as case studies. This selection represents good economic performance (Koto Mesjid) and poor economic performance (Tanjung Balik) after resettlement. More than 80% of villagers from each village were self-employed farmers involved in rubber plantations; this proportion increased further after resettlement. However, the difference in the income level between the two villages significantly increased after resettlement whereas prior to resettlement it had been quite small. Only 6% of residents of Tanjung Balik earned IDR4.5 million per month while 26% of residents of Koto Mesjid were at that level of income. This difference can be attributed to the introduction of aquaculture in Koto Mesjid: 36% of the residents are now involved in fish breeding as a secondary source of income whereas only 4% of the residents of Tanjung Balik are in involved in this industry. The entrepreneurship of Koto Mesjid residents has led to this economic success.

Bili-Bili Dam, Indonesia

The Bili-Bili Dam was constructed in the South Sulawesi Province of Indonesia. This project was funded by the OECF and the main purposes of the dam were to control flooding, to supply drinking water and to supply irrigation. As a result of construction, 2085 families were displaced with only cash compensation. These resettled families were free to choose their destination and they were given the opportunity to join a transmigration programme that was originally exclusive to those living in Java and Bali. Those who joined the programme went to newly developed resettlement areas in the Mamuju and Gowa Districts of Sulawesi Island, some hundreds of kilometres away from their homes. The relocation began in 1990 and was completed in 1997. After a few years, however, many of the individuals who had joined the transmigration programme returned to the areas closer to the reservoir and their original land. The major reason for this return was attachment to their original home (Rampisela, Solle, Said, & Fujikura, 2009).

Yoshida, Agnes, Solle, and Jayadi (2013) conducted a survey of 101 returnees to examine the role of the transmigration programme as an additional option in the relocation scheme. The survey revealed that those who had joined the transmigration programme were poor families that could not afford land or homes in the vicinity of the submerged land. The returnees found the conditions in the transmigration area too difficult and/or they wanted to live closer to their families. Those who had successfully saved money in the transmigration areas returned and bought land and homes closer to their original land. Some of the returnees gave to their children, or illegally sold, the land they had purchased with dam compensation money. Some of those who had not been able to save enough money in the transmigration areas nevertheless returned to the dam vicinity and are currently living with family or other relatives. Many of the second generation are still living in the transmigration areas because the living conditions have significantly improved and they have less of an emotional attachment to the original land.

Nam Ngum 1 Dam, Laos

The Nam Ngum 1 Dam was built in 1971 and was the first large dam project in Laos. The financiers of the construction included the World Bank and the Japanese and Dutch governments. In total 23 villages, with 570 households and 3242 people, were affected by this construction. The development plans began as early as 1957 and therefore no environmental impact assessment was conducted. Resettled individuals were less concerned with the social and environmental impact since they believed that the dam was important to the development of their country. Replacement land was provided for the resettled individuals and no cash compensation was provided.

Souksavath and Maekawa (2013) conducted a survey of 50 resettled individuals in each of two villages, Pakcheng and Phonhang, to compare livelihood conditions between the two villages. The survey revealed that the resettled individuals had had no choice other than to move into the designated resettlement villages. Both villages have continued the tradition of irrigated paddy rice cropping, as rice is the main food staple for those in the village. Although the distance between the two villages is only about 1 km, the disparity between them was significant. The average income per household per annum was USD2889 and USD1298 in Pakcheng and Phonhang, respectively. This disparity may be attributed to access to irrigation water and access to the main road. Packcheng is directly connected to the main road toward Vientiane and thus provides villagers with additional opportunities for employment. A distance of only 1 km significantly hinders opportunities for the villagers of Phonhang. They had no choice but to relocate and only the timing of the relocation determined their new living conditions.

Nam Theun 2 Dam, Laos

Construction of the Nam Theun 2 Dam was completed in 2008. The Nam Theun 2 Electricity Consortium conducted the entire development project and various international and private banks financed the projects. The principal shareholders include financial institutions from France, Thailand and Laos (Nam Theun 2 Power Company, 2012). This project has the capacity to generate 1070 MW of electricity, of which 93% is destined for Thailand and 7% for domestic consumption within Laos. The dam construction affected

6738 people of 1298 households in 17 villages. Compensation was basically replacement land, and each village generally moved to a new resettlement area; two villages were divided into two or three groups and merged with others in the resettlement areas.

Souksavath and Nakayama (2013) conducted a survey of 135 resettled individuals moved from four of the villages. The survey revealed that the government had offered cash compensation but none of the resettled individuals had chosen this because they had believed that they would be well supported in the resettlement areas through a land-based resettlement package. Most of the resettled individuals indicated that they were satisfied with their current living conditions and employment. Many were self-employed farmers both prior to and after the resettlement and many villages were located in the mountains and thus isolated from the market prior to the resettlement. The project provided about 1000 hectares of community forest to each village as a supplement because the resettled individuals were only given 0.66 ha of land regardless of the area of land they once had. This was due to the limited availability of land in the project command area. The resettlement packages thus failed to satisfy these resettled individuals. Because some of the resettlement areas are located close to the main road, some resettled individuals were able to increase their income by sending their products to the market.

Kotmale Dam, Sri Lanka

The Kotmale Dam was constructed as one of the five major headworks projects undertaken in the Mahaweli Development Project, which covers approximately 210,000 hectares of farmland. The government of Sweden financially assisted this project. All together, 3961 families were resettled due to entire or partial inundation of their land (3056 families) or increased earth-slips in their farmlands (905 families). Resettlement was conducted from the late 1970s to the mid-1980s. Compensation was land-based and two options were provided: 1722 resettled individuals selected 2.5 acres of dry land with irrigation water and 0.5 acres of home plot at the new Mahaweli development areas ('Mahaweli systems'); 1334 resettled individuals selected 2 acres of tea plots in the vicinity of the Kotmale Reservoir.

Manatunge and Takesada (2013) conducted surveys of 437 resettled individuals in 2005 and 2011. The majority of these resettled individuals wanted to move to the Mahaweli systems because they wanted to continue paddy cultivation. There were some who did not want to move to the Mahaweli systems because the climate there is harsher and they did not want to leave their traditional villages. Therefore, they selected the Kotmale Reservoir area as their destination. More than 95% of those who settled in Kotmale are satisfied with the resettlement option whereas about 70% of those who settled in the Mahaweli systems indicate that they are satisfied. This satisfaction has a direct relation with the increase of income levels and income stability. Where income stability increased, the resettled individuals were much more likely to be satisfied with their choice of resettlement. The major crop in the Mahaweli systems is rice, but world rice prices dropped sharply throughout the 1980s. This decreased stability for resettled individuals. Resettled individuals in Kotmale have not faced such a decline in the price of their tea and were thus able to maintain their income. Their main concern is productivity because they do not have enough capital to invest in further productivity.

Atatürk Dam, Turkey

The Atatürk Dam was built in the south-eastern region of Turkey to meet the country's increasing energy and water demands. Construction began in 1983 and was completed in

1992. The land and homes of approximately 55,300 people were fully or partially inundated; 1 town and 11 villages were fully inundated and 3 towns and 79 villages partly inundated. Out of 55,300 people, 19,264 resettled to new areas that were developed as a part of the government project. Others moved to Adıyaman and Kahta, which are close to the inundated settlement sites, or voluntarily moved to Adana and Mersin, some hundreds of kilometres away from their homes.

Akça, Fujikura, and Sabbağ conducted a survey with 99 resettled individuals living around the reservoir areas. The interviewees consisted of two groups: 33 well-off farmers with incomes over USD1000/month and 66 poor individuals with incomes below USD1000/month. All compensation provided for the interviewees by the government was in cash. The government compensated the farmers on the basis of their real estate tax statements, which showed much less than the market price. Moreover, the price of land in the newly developed resettlement areas increased because of land speculation. Even though the actual compensation was about 40% to 60% less than they expected, large landowners could easily afford new farmland and thus continue farming. However, many poor farmers could not afford farmland and became share-croppers or labourers. The resettled individuals were generally dissatisfied with the resettlement, including the well-off farmers, who missed the opportunity to visit their parents and relatives. As for the poor resettled individuals who had lost their farms, because they are landless they are sometimes regarded as refugees in the new resettlement areas, and this wounds their pride.

Kusaki and Sameura Dams, Japan

In 1962, the Japanese government introduced the Guidelines on Standards for Compensation for Losses Caused by Public Land Acquisition. In these guidelines, the government declared that loss of land and property must be compensated with financial payment. But because a preliminary survey of the Kusaki Dam construction project started in 1958, these guidelines were not applied to the compensation. The resettled individuals received money for their material loss based on the standards set in place by the dam developer. Some 221 resettled individuals representing 103 households were moved out of their villages. Several compensation measures were implemented, including land near the original area, support for driving licenses, and job training.

A feasibility study of the Sameura Dam began in 1958 and the construction was completed in 1975 after intense dispute and negotiation between the dam owner and resettled individuals. The total number of resettled individuals was 352. In most of the affected areas of one village, 141 out of 167 households moved to neighbouring towns and other cities. After the compensation standards for individual compensation were announced, many villagers began to leave their village.

Matsumoto, Mizuno, and Onagi (2013) interviewed 28 resettled individuals and 4 village officers who were involved with the Kusaki Dam construction and 5 resettled individuals and 4 village officers who were involved with the Sameura Dam construction. The survey reveals that it is important to ensure the transparency of compensation rules among villagers in order to achieve better community rehabilitation. Socio-economic changes affect the lives of resettled individuals such that compensation measures must take into consideration the rehabilitation of a community in the long term. Without compensation guaranteeing life reconstruction, many are likely to leave their homes, seeking a new environment in an urban area. Plans put in place by the village government are crucial for development after completion of dam construction and to avoid unnecessary conflicts and emotional strain amongst villagers.

Conclusions and discussions

Case studies on 10 large dam projects were conducted, as summarized in Table 1. Many similarities were found among these cases. No one case provided a 'perfect' resettlement scheme and there were both winners and losers among resettled individuals without exception. Some good practices were identified, including those case studies that indicate that a land-for-land compensation scheme should still be a major option for resettled individuals. Moreover, resettlement packages not based on farmland (i.e. agriculture) could be shown as alternative options for resettled individuals. This implies that a new paradigm should be explored, in addition to land-for-land strategies, so that more options with solid viability can be offered for the resettled individuals.

In four cases (Koto Panjang, Nam Ngum 1, Nam Theun 2 and Kotmale), compensation was mostly land-based. In three of these cases (Nam Ngum 1, Nam Theun 2 and Kotmale), timing and/or selection of the destination determined the economic situation of the villages. Major factors in determining the level of satisfaction experienced by resettled individuals were not related to conventional agricultural production from compensated land. The existence of a means of secondary income generation proved pivotal. Access to urban areas was found to be crucial in these four cases. Resettled individuals were able to find employment in non-agricultural sectors outside of the villages in both the Nam Ngum 1 and Kotmale cases. In each case, the residents had very good access to markets in which to sell their secondary (non-agricultural) products. The level of entrepreneurship expressed by the resettled individuals was found to be a most important factor when generating secondary income sources.

In the other six cases (Wonorejo, Saguling, Bili-Bili, Atatürk, Kusaki and Sameura), the compensation was primarily monetary (i.e. cash compensation). Impoverishment, which the World Bank and the OECD have indicated is a concern with cash-only compensation, was observed in these six cases. In fact, the amount of compensation received for livelihood rehabilitation was completely insufficient in some cases, including Saguling, Bili-Bili, and Atatürk. Small landowners and landless farmers suffered many hardships as disparity among resettled individuals became larger. However, those who obtained adequate compensation managed to secure a better livelihood regardless of their choice to continue farming or take another job. In the Bili-Bili case, many of the large landowners indicated that they had abandoned farming and moved to urban areas to secure a better livelihood. In a country like contemporary Indonesia, or Japan in the early 1960s, land-for-land policies have caused farmers to remain relatively poor when compared to city dwellers. Those not

Table 1.   Dams surveyed.

Country	Dam	Resettlement	
		Period	Actual compensation
Indonesia	Bili-Bili	1990s	Cash
	Koto Panjang	1980s and 1990s	Land and cash
	Saguling	1980s	Cash
	Wonorejo	1990s	Cash
Laos	Nam Ngum 1	1960s	Land
	Nam Theun 2	2000s	Land
Sri Lanka	Kotmale	1970s and 1980s	Land
Turkey	Atatürk	1980s	Cash
Japan	Kusaki	1960s	Cash
	Sameura	1970s	Cash

engaged in agriculture tended to benefit substantially from the country's rapid economic development. Poor famers did not necessarily want to remain poor farmers after resettlement. Obtaining cash compensation, which allowed for a job rather than farming, was an attractive option for the farmers. Industrial development around resettlement areas also provided many (non-agricultural) opportunities.

Indonesia's national transmigration programme served as an effective safety net in the Wonorejo and Bili-Bili cases. Joining the transmigration programme was an additional benefit offered by the government, in particular in the Bili-Bili case. Resettled individuals seem to have suffered all type of adversities at the beginning of their resettlement due to insufficient infrastructure in the resettlement areas. However, they generally moved to better-off areas after relocation. Some of the first generation of Bili-Bili resettled individuals nevertheless returned from the transmigration sites to areas near their original homes. Their attachment to their original homes was very strong in many cases. This was observed not only in Bili-Bili, but also in Saguling, Kotmale and Atatürk. The emotional impact of resettlement (of the first generation in particular) should be taken into consideration when a resettlement plan is developed.

Another important factor is access to education and employment opportunities for the second and subsequent generations. In all the cases examined, resettled individuals place a high priority on the future of their children when selecting resettlement options. Thanks to both the economic development of the country and provision of public facilities by dam owners in the resettlement areas, the educational facilities have been significantly improved in most instances, except in Kotmale.

The Saguling case suggests that resettled individuals and their descendants many not have great access to secondary development. Promotion of secondary development should be accompanied by vocational training, technical assistance and the provision of capital. Allowing priorities for resettled individuals and some employment opportunities, such as work in the public sector (e.g. local municipal governments) may also be considered as part of the compensation package. Efforts in these new directions, namely from agricultural to non-agricultural sectors, should be taken into serious consideration, while the conventional land-for-land compensation scheme is still more desired in some countries. There are now some nations equipped with instruments to provide resettled individuals a better livelihood after relocation in sectors not based on the productivity of farmland, as demonstrated in the two Japanese cases in the 1960s and early 1970s when Japan experienced very rapid economic development.

Acknowledgements

This research was funded by the Mitsui & Co., Ltd, Environment Fund, the Foundation of River & Watershed Environment Management and KAKENHI (18310033 and 24310189).

References

Akça, E., Fujikura, R., & Sabbağ, C. (2013). Atatürk Dam resettlement process: Increased disparity resulting from insufficient financial compensation. *International Journal of Water Resources Development, 29*, 101–108.

Development Assistance Committee (1992). *Guidelines for aid agencies on involuntary displacement and resettlement in development projects* (*Guidelines on Aid and Environment*, No. 3). Paris: OECD.

Fujikura, R., Nakayama, M., & Takesada, N. (2012). Lessons from resettlement caused by large dam projects: Case studies from Japan, Indonesia and Sri Lanka. *International Journal of Water Resources Development, 25*, 407–418.

International Rivers (2012). *China's global role in dam building*. Retrieved from http://www. internationalrivers.org/it/node/2312

Karimi, S., & Taifur, W. D. (2013). Resettlement and development: A survey of two of Indonesia's Koto Panjang resettlement villages. *International Journal of Water Resources Development, 29,* 35–49.

Manatunge, J., & Takesada, N. (2013). Long-term perceptions of project-affected persons: A case study of Kotmale Dam in Sri Lanka. *International Journal of Water Resources Development, 29,* 87–100.

Manatunge, J., Takesada, N., Miyata, A., & Nakayama, M. (2009). Livelihood rebuilding of dam-affected communities: Case studies from Sri Lanka and Indonesia. *International Journal of Water Resources Development, 25,* 479–489.

Matsumoto, K., Mizuno, Y., & Onagi, E. (2013). The long-term implications of compensation schemes for community rehabilitation: The Kusaki and Sameura dam projects in Japan. *International Journal of Water Resources Development, 29,* 109–119.

Nam Theun 2 Power Company (2012). Principal Shareholders and Roles in the Project. Retrieved from http://www.namtheun2.com/index.php?option=com_content&view=article&id=64& Itemid=61

New Energy Foundation (2010). *Report on hydropower toward low-carbon society* [in Japanese]. Tokyo: Author.

Rampisela, A. D., Solle, M., Said, A., & Fujikura, R. (2009). Effects of construction of the Bili-Bili Dam (Indonesia) on living conditions of former residents and their patterns of resettlement and return. *International Journal of Water Resources Development, 25,* 467–477.

Sisinggih, D., Wahyuni, S., & Juwono, P. T. (2013). Assessing the perception of former residents on the implications of the resettlement program of Wonorejo Dam, Tulungagung, Indonesia. *International Journal of Water Resources Development, 29,* 14–24.

Souksavath, B., & Maekawa, M. (2013). The livelihood reconstruction of resettlers from the Nam Ngum 1 hydropower project in Laos. *International Journal of Water Resources Development, 29,* 59–70.

Souksavath, B., & Nakayama, M. (2013). Reconstruction of the livelihood of resettlers from the Nam Theun 2 hydropower project in Laos. *International Journal of Water Resources Development, 29,* 71–86.

Sunardi, Gunawan, B., Manatunge, J., & Pratiwi, F. D. (2013). Livelihood status of the resettlers affected by the Saguling Dam project, 25 years after inundation. *International Journal of Water Resources Development, 29,* 25–34.

Suwartapradja, O. S., Arifin, T., Kanum, A., Ansor, & Djumari (1985). *Pemantauan sosial-ekonomi budaya penduduk pindahan dari bawah ke atas genangan PLTA Saguling* [Monitoring on socioeconomic and cultural of the displaced people]. Bandung, Indonesia: Pusat Penelitian Sumber Daya Alam dan Lingkungan, Universitas Padjadjaran.

Yoshida, H., Agnes, R. D., Solle, M., & Jayadi, M. (2013). A long-term evaluation of families affected by the Bili-Bili Dam development resettlement project in South Sulawesi, Indonesia. *International Journal of Water Resources Development, 29,* 50–58.

World Bank (2004). *Involuntary resettlement sourcebook: Planning and implementation in development projects*. Washington, DC: World Bank.

World Commission on Dams (2000). *Dams and development*. London: Earthscan.

The resettlement programme of the Wonorejo Dam project in Tulungagung, Indonesia: the perceptions of former residents

Dian Sisinggih[a], Sri Wahyuni[b] and Pitojo Tri Juwono[a]

[a]Department of Water Resources Engineering, Faculty of Engineering, University of Brawijaya, Malang, Indonesia; [b]Department of Civil Engineering, Faculty of Engineering, University of Jember, Indonesia

The Wonorejo Dam project involuntarily relocated many families from the dam site. These resettled individuals opted to move into surrounding villages rather than to follow the transmigration scheme put in place that would have taken them beyond Java Island. Although the former residents were moved involuntarily, many of them are found to be content with their current situation and conditions. The findings of this study may help appropriate authorities enhance their social responsibility and evaluate their respective resettlement programmes.

Introduction

This study attempted a more detailed investigation of the perceptions of former residents of the Wonorejo Reservoir area on Java Island of Indonesia by considering residents who opted to move upstream and downstream of the reservoir. This also included the second generation of former residents. Resettled individuals who had not chosen the transmigration scheme but moved to surrounding villages (including the upstream area) were also considered as respondents in this study.

The Wonorejo Multipurpose Dam was the 20th dam constructed in the Brantas River basin (Sinaro, 2007). The project site is located in Wonorejo Village, Pagerwojo sub-district, Tulungagung Regency, East Java Province, Indonesia. This region is about 130 km south-west of Surabaya, the capital city of East Java Province, and about 15 km west of the Tulungagung city centre. The dam was built in Gondang River, about 400 m downstream of the confluence of the Bodeng and Wangi Rivers.

The main objective of the project was to improve the low economic status in the Tulungagung area by increasing agricultural production through irrigation. Meanwhile, severe droughts affected the Brantas Basin both in 1982 and in 1987. The city of Surabaya suffered from an acute shortage of water because its main source was the Brantas River. At this point, project developers determined that the project should also include urban water supply for Surabaya. The Asian Development Bank (ADB) financially supported this project from 1980 to 1986.

11

The Wonorejo Dam has a height of 100 m and inundates an area of about 3.85 km^2 at its highest water level. According to a survey in 2008, the effective storage of this reservoir is 99,040,000 m^3.

The Wonorejo Dam currently delivers the following services: (1) water provision for domestic use and for industry downstream of the Brantas River to Surabaya and its vicinity (in the past, users faced great difficulty meeting their water requirements for basic needs and economic activities); (2) flood control for the Gondang River; (3) hydro-electric power generation (6300 kW) – additionally, a micro-hydro electric power plant (236 kW) has been established utilizing water outflow from a hollow jet valve (this energy production is sufficient to meet the demands of the facility at the Wonorejo Dam); (4) tourism support; and (5) inland fisheries.

The Wonorejo Multipurpose Dam project is part of a comprehensive development scheme for the Brantas River basin (Nippon Koei, 1992). The project received technical assistance from Japan's Overseas Technical Cooperation Agency (now the Japan International Cooperation Agency) in 1973.

Dam construction was completed in 2000. The financial support for this project was provided by the Indonesian government, along with a loan from the Overseas Economic Cooperation Fund of the Japanese government. Indonesia's vice-president inaugurated the dam and its facilities on 20 June 2001.

Resettlement programme

Dam development projects often require the involuntary resettlement of a number of people, depending on the extent of the land that is to be submerged (Fujikura, Nakayama, & Takesada, 2009). In the case of the Wonorejo Dam project, those residing in the reservoir area became subject to an involuntary resettlement programme.

The Wonorejo Dam is located in the village of Wonorejo, which consisted of six sub-villages: Suruh, Bendungan, Boro, Wates, Jeruk and Suwaloh. According to the survey report in 1980, 7144 individuals (in 1414 households) lived in the village. Land use prior to the dam construction was as follows: paddy fields occupied an area of 189 ha (105 ha irrigated, 84 ha non-irrigated); dry land included about 306 ha; residential/ housing, 125 ha; forest, about 1250 ha; other land uses, approximately 4.3 ha. Land acquisition began in 1982/1983 when the government executed its plan to accelerate the project.

Based on a topographical survey, the area of Wonorejo Reservoir was predicted to cover 210 ha. In May 1981, to obtain more accurate data, a second survey was conducted within the area to be inundated. This survey indicated that the submerged area would be approximately 210 ha and affect 668 households. Another 909 households on 450 ha of land lived on terrain that would be isolated by the reservoir (IPB, 1985).

There were three scenarios for resettlement prepared by the Government.

1. Resettlers could be moved to a new area outside of Java Island (preferably the southern portion of Sumatra Province). In this scheme, applicable to all 1414 affected households, resettlers needed to join the transmigration programme then being promoted by the government. This programme was intended to move landless people from densely populated areas within Indonesia to less populous areas of the territory. In practice, this meant that individuals from the islands of Java, Bali and Madura would be moved to Papua, Kalimantan, Sumatra and Sulawesi.

2. A partial resettlement programme targeted only those living in the area to be submerged (668 households). This scenario was conducted using the transmigration programme or relocation to nearby villages.
3. There was a swamp reclamation project in the southern part of Tulungagung. This project had been intended to absorb many resettlers. It later turned out that farmland newly reclaimed by the project would be ready by 1985/1986 and could accommodate only 668 households, much fewer than initially estimated. Therefore, the government decided not to implement this swamp reclamation project. As a result, the resettlement programme was composed of a combination of the first two scenarios (Nippon Koei, 2002).

According to the survey carried out by the authority in charge of the project, about 475 households decided to join the transmigration scheme, about 356 moved to surrounding villages, and 164 chose to resettle upstream after receiving their compensation. This latter group owned dry and agricultural land in addition to still having relatives in the project area. The upstream area included remote villages with poor access to roads and fresh water, no sanitation, and no electricity. These conditions were certainly precarious.

The resettlement programme attempted to at the very least sustain or improve the living standards of resettled individuals. Furthermore, the project constructed infrastructure at the resettlement sites but also recognized that the growing number of the settlers moving upstream meant an increased environmental load, leading to greater environmental deterioration.

It is important to examine the satisfaction of former residents in order to understand whether the implementation of this project was successful and also to evaluate the sustainability of their positive perceptions. A condition of Japan's Official Development Assistance loan included an interview with 20 resettled residents in October 2004 to investigate the current condition of various public facilities built related to the project and to obtain their opinion concerning the resettlement process and their current lifestyles within the village (Okada, 2004). A summary of the interview results is as follows.

1. The land acquisition and resettlement process involved no disputes with residents and proceeded smoothly from beginning to end.
2. Roads, school buildings, health centres and other public facilities were constructed, and resettled residents were satisfied with their living environment.
3. Although resettled individuals had suffered no economic hardships in the past as a result of farming, transporting materials or supplying labour during construction of the dam, they had found it more difficult to find work since the project was completed and had become anxious about their potential income. In this respect, it would be beneficial for the government to provide vocational training and job opportunities.

As these comments show, while the residents were not negative about their resettlement, they indicated that more thought should have been given to their situation several years down the road. Their hope was that the government would provide some form of vocational training as a means of improving future living standards. Accordingly, plans to support the livelihood of residents after resettlement should have been investigated and implemented.

Since the perception of former residents is important for sustainable project management, this kind of study is required to evaluate and to monitor the long-term consequences of resettlement programmes (ADB, 1999; IFC, 2002). However, the previous study made no detailed description concerning the selection of interviewees and their spatial distribution. This study has attempted to investigate in more detail the perceptions of the former residents of the Wonorejo Reservoir area.

Research objective and methodology

This study assesses the level of satisfaction of former Wonorejo residents who followed different resettlement programmes. Further, it identifies the opinions of the second generation concerning the involuntary resettlement process. The findings of this study are important for authorities and related institutions to evaluate resettlement programmes and to enhance their future schemes based on this community approach.

The field investigation and interviews undertaken to map the perceptions of individuals exposed to an involuntary resettlement programme focused on former residents of the dam construction area. The primary points addressed in the interviews included the community's response to the resettlement programme and their adaptive capacity to a new environment.

Because it would be possible for certain factors to influence those who completed the questionnaire, respondents were randomly selected as a heterogeneous crowd where different groups of people represented different age groups and both sexes.

A key distinction considered in the selection of interviewees was between:

1. those who held an independent household before resettlement and who made the decision at the time of resettlement (they were considered first-generation); and
2. those who became independent at the time of resettlement or after – this also included those who did not make their own choice regarding resettlement and obeyed the decision made by their parents (this group of people were considered second-generation).

The questionnaire included a pre-test aimed at appraising certain factors within a small number of interviewees before the study was conducted in the field. Such preliminary assessment is useful for researchers in identifying the strengths and weaknesses of the questionnaire, including its reliability and validity. Another pre-test approach is to ask people how they interpreted the different concepts in the survey (Saris & Gallhofer, 2007).

Pre-test and field investigations were carried out in July 2011, while delivery of the questionnaires and an additional field survey were completed by August that year. Another informal interview with local authorities and subsequent field observations were carried out to gather as much information as possible.

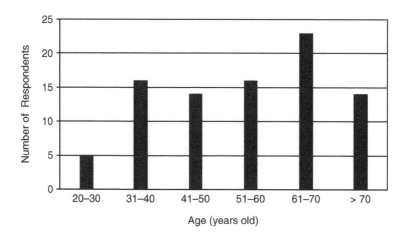

Figure 1.   Distribution of age of respondents.

The number of respondents in the study was 88. The questionnaire topics were classified into nine groups based on occupation/income, land ownership and farming activities, fishing, property, convenience in daily life, their children's educational opportunities, health, social community, and general satisfaction. Other factors included compensation and administrative problems.

Findings and discussion

The 88 respondents were randomly selected, based on the pre-survey, to draw from the spatially distributed surrounding villages chosen to be resettlement areas. The second generation was also involved, to obtain their perspectives on the progress of the resettlement scheme in general.

Figure 1 indicates that the age of the respondents varied from teenagers to the elderly. The younger respondents were largely from the second generation, while the older respondents were from the first generation. It was quite difficult to track down both of these generations since they were often living sparsely and had selected their own resettlement areas. Most of the members of the first generation had been farmers prior to the resettlement; only few of them had changed their occupation by becoming businessmen or labourers. The second generation mostly engaged in the same activities as their parents.

Because the second generation often lived near their parents, they were quite easy to locate. In some cases, it was difficult to obtain the thoughts of the second generation because they were too young to be considered in this study. Table 1 shows the distribution of respondents according to their chosen resettlement area and surrounding villages. For the purposes of this figure, the label "other" designates the villages located downstream of the reservoir, closer to urban areas.

The respondents did not all participate in the resettlement scheme at the same time, as is shown in Table 2. In the initial stages, resettlement was conducted gradually and afforded first priority to those living in the submerged area; second priority was then given to residents near the construction site. Since there were financial problems with the dam construction when donor countries suspended their support, first-priority relocations only started in 1982/83 and extended until 1992/93. In 1994/95, the government committed national resources to continue the construction of the Wonorejo Dam. The resettlement programme was re-initiated in 1994/95 and 1995/96. Accordingly, the respondents were identified as first-priority (44 respondents) and second-priority (44 respondents).

Some main indicators of public perception are summarized in Table 3. Before accepting the resettlement programme, many of the respondents did not own land to

Table 1. Number of respondents among sub-villages.

Sub-village	Boro	Wates	Dawuhan	Suwaloh	Other
First generation	35	18	3	1	13
Second generation	16	0	0	0	2

Table 2. Number of respondents per period of resettlement.

Year	1982/83	1983/84	1984/85	1992/93	1994/95	1995/96
Respondents	36	1	1	6	41	3

Table 3. Summary of number and percentage of respondents concerning conditions before and after the resettlement program according to respondent classification.

Main indicator	Relocation area				Relocation period				Generation			
	Upstream		Downstream		1st stage		2nd stage		1st generation		2nd generation	
	Number	%	Number	%	Number	%	Number	%	Number	%	Number	%
A. Income stability												
Better	24	33	0	0	7	16	17	39	14	20	11	61
Worse	30	41	6	40	17	39	14	32	34	49	1	6
Other	10	14	2	13	6	14	6	14	10	14	2	11
No answer	9	12	7	47	14	32	7	16	12	17	4	22
B. Land ownership												
Increase	54	74	13	87	40	91	27	61	58	83	9	50
Decrease	17	23	2	13	3	7	16	36	10	14	9	50
Other	0	0	0	0	0	0	0	0	0	0	0	0
No answer	2	3	0	0	1	2	1	2	2	3	0	0
C. Property												
Increase	37	51	4	27	19	43	22	50	27	39	14	78
No change	23	32	5	33	16	36	12	27	25	36	2	11
Decrease	11	15	6	40	9	20	8	18	16	23	1	6
No answer	2	3	0	0	0	0	2	5	2	3	1	6
D. Social community												
Better	63	86	14	93	39	89	39	89	61	87	16	89
Worse	8	11	1	7	5	11	4	9	7	10	2	11
Other	1	1	0	0	0	0	1	2	1	1	0	0
No answer	1	1	0	0	0	0	0	0	1	1	0	0
E. Compensation scheme												
Better	45	62	10	67	30	68	26	59	47	67	7	39
Worse	6	8	4	27	8	18	2	5	7	10	3	17
Other	21	29	1	7	6	14	16	36	14	20	8	44
No answer	1	1	0	0	0	0	0	0	2	3	0	0

Table 3 – continued

Main indicator	Relocation area				Relocation period				Generation			
	Upstream		Downstream		1st stage		2nd stage		1st generation		2nd generation	
	Number	%	Number	%	Number	%	Number	%	Number	%	Number	%
F. General satisfaction												
Better	68	93	15	100	43	98	27	61	67	96	16	89
Worse	0	0	0	0	0	0	0	0	0	0	0	0
Other	0	0	0	0	0	0	0	0	0	0	0	0
No answer	5	7	0	0	1	2	17	39	3	4	2	11
Total number of respondents	73		15		44		44		70		18	

engage in agriculture. Fifty-four (74%) of the respondents living upstream and 13 (87%) of those living downstream indicated that they presently own more land. According to the resettlement priority that an individual was associated with, there may have been more or less land available. Similarly, 40 (91%) of the respondents from the first priority and 27 (61%) of the respondents from the second priority reported owning more land than previously. The remaining resettled individuals struggled to find adequate land for agriculture.

Most respondents experienced instability with their income after entering the resettlement programme, though 17 respondents from the second priority of resettled individuals (39%) expressed that their income was stable. This was thanks to better compensation and a better political situation within this period. The instability associated with the first-priority resettled individuals was linked to insufficient skills and age; competition for new jobs was difficult when they had only worked in the agricultural sector. Except for income stability, there were almost no reported differences in the livelihood condition between respondents living in the upstream or downstream areas or between those who resettled in different years. There were advantages and disadvantages for both the upstream and downstream areas. Those who moved to the downstream area of the reservoir had access to better technology and good public facilities; however, their economic burden was heavier than those living in the upstream area. As a consequence, six (40%) of these respondents indicated that that their property was reduced. In contrast, those living in the upstream area of the reservoir obtained additional income by temporarily utilizing forested land for agriculture. They now have enough land for cultivation, although the quality of land is worse, compared to what they owned in the past. This was one of the reasons that those living upstream indicated that their income was not stable.

Resettlement programme

This study found that former residents had strong emotional reasons for choosing to remain in the surrounding villages. They had been introduced to the options of transmigration or moving to nearby villages. The government also promised those in the transmigration scheme more land and greater cash compensation. Despite available options, they chose to move to nearby villages for emotional reasons. Specifically, 63 respondents (59%) had wanted to stay near the reservoir area, 25 (23%) still occupied land that was not submerged, 7 (7%) were not confident with the government's promise and their own capacity to survive in the transmigration area, and 12 (11%) gave other reasons.

During the land acquisition processes, those whose land was either fully or partially submerged received assistance from local government officers and community leaders in negotiating cash compensation. Only those who had partially submerged land indicated that they did not receive enough cash compensation; this situation rendered such respondents landless. According to their answers, 72 respondents (33%) indicated that after receiving compensation cash they had almost enough to set up their new house, and 64 respondents (29%) indicated that they had enough for a daily meal. Only a few of the respondents indicated that their cash compensation contributed to their savings. Hence, the results seem to suggest that cash compensation alone was not enough for the resettled individuals (particularly for those whose land was only partially submerged) and they were unable to reconstruct their lives.

Transmigration scheme

According to the respondents' knowledge of the condition of their neighbourhood, 52 respondents (59%) reported a success story in terms of their achievement in the new area while 3 (3%) believed that they did not succeed in the transmigration area. The remaining 33 respondents (38%) reported not having enough information on their neighbourhood. Sometimes reunions take place and transmigrated people visit their relatives in the surrounding villages. This possibility might change the perception of the transmigration scheme as an option in the resettlement programme, but it could not change their emotional reasons for remaining nearby.

Respondents' views on the resettlement programme

To date, almost all of the respondents were satisfied with the education facilities provided in the resettlement programme. Also, in contrast to the first generation, the second generation suggested that their income was more stable in their present setting. This means that the first generation, especially those who live downstream, continued to struggle to adapt to this new area. They also experienced difficulty as job seekers. Most of the respondents expressed that their current situation posed greater difficulty in obtaining suitable jobs. This is reasonable since farming was their major former job and most current jobs require greater skills and younger staff. Despite the first generation's difficulties in adaptation, the second generation easily adapted to the new area. Regarding their housing condition, all of the residents indicated that their present home was better than their previous one.

General satisfaction with the resettlement programme

The majority of respondents indicated that they are satisfied with their living conditions. Their houses are larger than previously and the social and public facilities are much better. The respondents who elected to live in the downstream area have better education facilities as well as more opportunities to obtain higher education for their children. Also, they have better and easier access roads to all public facilities (e.g., hospitals, banks, the city centre, and others). In contrast, respondents living in the upstream area are very much isolated. On the other hand, they earn a more stable income because they have more opportunities to work in the agricultural sector. In general, both were satisfied with their advantages and disadvantages. Regarding community adaptation, the respondents reported no major difficulty living in the resettled area; they were adapting easily to their new neighbourhoods and environment.

Conclusions and suggestions

According to the respondents' opinions, those living in the upstream areas were more satisfied with their living conditions, compared to those living in the downstream areas. The upstream residents apparently have access to a more stable income. The interview process revealed that those living upstream temporarily have the opportunity to utilize and cultivate forestry. So long as there are no negative impacts for either reservoir or forest, this mutual condition should be maintained by the government, via the forestry ministry, to maintain the livelihood condition of residents living upstream. This is also part of sustainable project management. Unlike those living in the upstream area, those living downstream found it difficult to secure a stable income. This opinion emerged more strongly from the first generation than from the second.

Concerning the compensation process and negotiations, the majority of respondents felt that they were fair enough, despite the fact that their expectations were not completely fulfilled. Most of the resettled individuals spent their cash compensation to purchase new land and to establish a new home. Because they had moved to a remote area, they could get more land compared to what they previously owned. In addition, they also owned more precious belongings, except those who moved to the downstream area. Residents living in the downstream area had higher economic loads that those living upstream and as a consequence they had been forced to spend much of their cash compensation.

In general, the respondents' perception of the implementation programme was positive; however, this could still change, particularly with the group living in the upstream area. Authorities should seriously monitor this perception. In in-depth interviews, those in the upstream area reported that they are isolated from others due to the degradation of the access roads. Continued neglect of this situation may affect their economic activities and daily life.

In order to make a comprehensive assessment of the perceptions of former residents of the Wonorejo Dam project, it is recommended that further study incorporate the perceptions of those who were resettled using the transmigration scheme, since there are success stories from these individuals also.

Acknowledgements

The Mitsui & Co., Ltd, Environment Fund financially supported the research carried out within this study. This study was also partially supported by KAKENHI (24310189). Core Research for Evolutional Science and Technology of the Japan Science and Technology Corporation also funded this study. The authors would like to thank Jasa Tirta for providing secondary data and reports. Gratitude and sincere appreciation are also extended to Mr Naruhiko Takesada for draft questionnaires. Thanks also to those individuals and institutions who were involved in this study.

References

ADB (Asian Development Bank) (1999). *Buku Panduan Tentang Pemukiman Kembali - Suatu Petunjuk Praktis* [Handbook of resettlement: a practical guide]. Manila: Author.

Fujikura, R., Nakayama, M., & Takesada, N. (2009). Lessons from resettlement caused by large dam projects: case studies in Japan, Indonesia and Sri Lanka. *International Journal of Water Resources Development, 25*, 407–418.

IFC (International Finance Corporation) (2002). *Handbook for preparing a resettlement action plan.* Washington, DC: Author.

IPB (Institut Pertanian Bogor) (1985). *Draft final report - Analisa Dampak Lingkungan pada Proyek Tulungagung: Pembangunan Waduk Wonorejo* [Draft final report on environmental impact assessment of Tulungagung Project: the development of Wonorejo Dam]. Bogor: Author.

Nippon Koei (1992). *Definitive plan study report on Wonorejo Multipurpose Dam projects.* Jakarta: Directorate General of Water Resources Development, Ministry of Public Works.

Nippon Koei (2002). *Completion report, Volume II: supporting report package-1B infrastructures at resettlement site.* Jakarta: Directorate General of Water Resources Development, Ministry of Public Works.

Okada, T. (2004). *Wonorejo Multipurpose Dam Construction Project*, Japan International Cooperation Agency. Retrieved from http://www.jica.go.jp/english/our_work/evaluation/oda_loan/post/2005/pdf/2-05_full.pdf

Saris, W. E., & Gallhofer, I. N. (2007). *Design, evaluation, and analysis of questionnaires for survey research*. Hoboken: John Wiley & Sons.

Sinaro, R. (2007). *Menyimak Bendungan di Indonesia (1910–2006)* [Review on Indonesian Dams (1910–2006)]. Jakarta: Bentara Adhi Cipta.

Livelihood status of resettlers affected by the Saguling Dam project, 25 years after inundation

Sunardi[a,b], Budhi Gunawan[b], Jagath Manatunge[c] and Fifi Dwi Pratiwi[b]

[a]Department of Biology, Faculty of Mathematics and Natural Sciences, Universitas Padjadjaran, Sumedang, Indonesia; [b]Institute of Ecology, Universitas Padjadjaran, Bandung, Indonesia; [c]Department of Civil Engineering, University of Moratuwa, Sri Lanka

A study of the effects of the Saguling Dam project has been conducted. This paper attempts to examine the long-term effects of the dam construction on the livelihoods of the displaced people, paying special attention to any effects caused by inequality of access to resettlement schemes. The study results indicate that the majority of the resettlers perceived their livelihoods as being better after their resettlement. However, loss of jobs or conversion to less preferable or beneficial occupations caused by the project has affected their satisfaction level. In addition, inequality of access to options of the resettlement scheme has caused differences in socio-economic status among the resettlers. Furthermore, in the long term, the option has also failed to indemnify resettlers from lost livelihoods due to environmental and socio-economic constraints. For future resettlement programmes, the authors propose that policy makers should employ analysis instruments which can precisely predict long-run impacts, while local backgrounds and dynamics are important to be considered to secure the success of resettlement programmes.

Introduction

In West Java Province, Indonesia, Saguling Dam was constructed about 25 years ago with the main purpose of provision of hydroelectric power to heavily populated Java and Bali. The project displaced 3038 families from the inundated area and affected 7626 families that lived in non-inundated areas but had land and sources of income in the inundated area (PLN, 1989). It has been acknowledged by researchers that many of the most challenging socio-economic impacts of dam construction relate to the migration and resettlement of people near the dam site or in the catchment area (Bartolome, de Wit, Mander, & Nagraj, 2000; Cernea, 2003; Egre & Senecal, 2003).

In view of the significant number of people displaced by the project, the government provided a resettlement programme. Several schemes were offered: (a) transmigration to islands outside Java; (b) local transmigration (within West Java); and (c) decision by the resettlers as to where they would move. A few additional alternatives, such as estate work, construction and agri-aquaculture were also provided as options by the government. Only 3.9% of the displaced people followed the first two schemes; the rest chose to relocate near the reservoir (Suwartapradja, Arifin, Kanum, Ansor, & Djumari, 1985). Also, some of the resettlers who chose to transmigrate to outside Java and some of those who had moved out based on their own choice returned to the area surrounding the lake for various reasons. Consequently, the population density around the lake increased.

In such a case, a local resettlement scheme would perhaps assist in meeting the goals of large-scale resettlement. As a result of the environmental impact assessment recommendations on resettlement (IOE, 1979; Soemarwoto, 1990), the PLN (*Perusahaan Listrik Negara*: State Electricity Company) implemented a local resettlement scheme unique to Indonesia. Local resettlement supplemented by livelihood-rebuilding opportunities in aquaculture, tourism and small-scale industry, using the electricity provided by Saguling, was planned. However, the "aquaculture resettlement option" was a priority of PLN/IOE/ICLARM/World Bank efforts (Soemarwoto, 1990; Sutandar, Costa-Pierce, Iskandar, & Hadikusumah, 1990), and it would play the most important role in helping resettlers rebuild their livelihoods quickly after the inundation.

Development of cage aquaculture and capture fisheries was, then, initiated to provide rural jobs and to maximize all possible productive uses of the new water resources. The use of floating net cage (FNC) aquaculture was proposed because it was deemed compatible with engineering forecasts of reservoir operations and draw-downs, and also because the reservoir had many deep, sheltered bays very suitable for cage aquaculture (Gunawan, 1992; Manatunge, Contreras-Moreno, & Nakayama, 2001; Nakayama, 1998). A total of 1500 families were the target of the cage aquaculture training programme conducted by IOE/ICLARM in cooperation with the West Java Fisheries Agency, under the direction of the PLN (IOE-UNPAD & ICLARM, 1989). However, there is no record of the exact number of people involved in that programme. The southern sector of Saguling is now the predominant centre of the Saguling FNC aquaculture industry. The success of the floating net cage system in this area is due not only to better water quality factors, but also to socio-economic and infrastructural factors. On any account, inequality of access to resources and opportunities among the resettlers was noticed from the onset, particularly dealing with the cage aquaculture that ultimately affected their socio-economic status.

This paper examines the long-term effects of Saguling Dam construction on the livelihood of the displaced people, paying special attention to the effects of inequality of access to FNC aquaculture between the two villages.

Research methodology

Research sites

This research was conducted at two villages near areas which were inundated by the Saguling Dam project in 1985: Bongas (in Cililin Subdistrict) and Sarinagen (in Cipongkor Subdistrict). Bongas lies in the southern section of Saguling Lake, which in general has better water quality compared to other parts of the lake, including Sarinagen in the northern section (see Figure 1). Shortly after inundation, the cage aquaculture development programme was introduced as a part of local resettlement schemes which aimed to provide employment for resettled people. Bongas was selected as the site of the cage aquaculture programme implementation and as the centre of training and supporting activities. It is proposed that the success of the cage aquaculture programme in Bongas was due to better water quality and socio-economic and infrastructural support. Recently the practice of FNC aquaculture in the Bongas area is still common, while in the Sarinagen area it is rare.

23

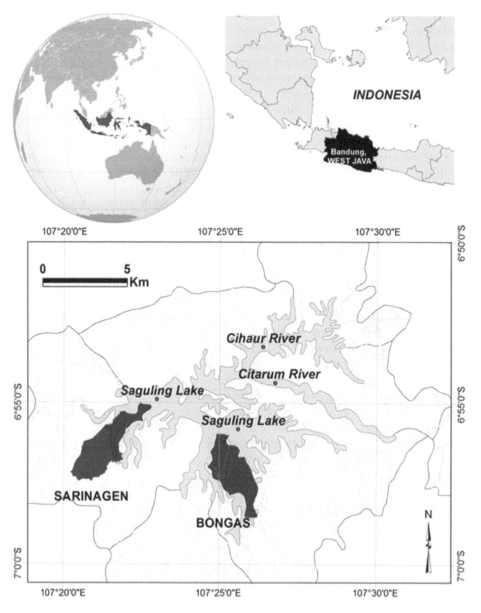

Figure 1.   The villages of Bongas and Sarinagen (shaded areas).

Data collection and analysis

The study employed a descriptive approach, intended to explain certain characteristics occurring in the resettlers' livehoods both in Bongas and Sarinagen. Both quantitative and qualitative data were used. A survey was carried out in 2011 by interviewing a number of selected heads of household to reveal their socio-economic status and perceptions. Observation was also conducted and secondary data were collected to describe the general condition of the studied villages.

The interviewed heads of household were those who had been at least 17 (or married) at the time of dam construction. It was assumed that they had strong perceptions about any impacts of the dam construction in the past, and could provide any relevant information clearly. The respondents were 147 resettlers, 97 from Bongas and 50 from Sarinagen. As the population of the resettlers was unknown, non-probability sampling was employed. Targeted respondents were traced through a list provided by the village administrative offices.

Research findings

Income/occupational aspect

With regard to their occupation, primary attention was on any changes of employment caused by dam development. Because people were involuntary displaced from their original settlements to unplanned sites, some socio-economic hardships resulted, including changes in employment and income-generating opportunities. The construction of Saguling Dam had forced their jobs to change. The number of self-employed farmers decreased drastically, both in Bongas and Sarinagen; in contrast, the number of share-croppers increased. In addition, the survey found that the number of unemployed among the resettlers had increased markedly, both in Bongas and in Sarinagen. Certainly many of the resettlers who were interviewed were no longer young; however, in rural areas it is common for old men to continue to work for a living. At present, the numbers of people who are employed in public offices and the private sector, and those employed as labourers, are stable. Meanwhile, other sectors such as business, entrepreneurs and services employ a considerable proportion among the resettlers (23% to 30%). The tourism and estate sectors, which were developed by the local government and the PLN, were not successful (see Table 1).

FNC aquaculture ownership

At the beginning of the programme, the implementation of FNC aquaculture development was aimed to absorb persons who had become unemployed due to dam construction. The Bongas area became the central site, and is the predominant centre for FNC aquaculture even at present. A few years after its development, however, the FNC population in the Saguling Dam area declined continually as a consequence of fish diseases, poor water quality, and the national economic crisis in Indonesia in 1998. Most of the resettlers in both villages were not able to reconstruct their FNC aquaculture because of limited access to capital and assistance. Recent data show that the resettlers

Table 1. Resettlers' occupations before resettlement and at present in Bongas and Sarinagen. (Multiple answers were permitted).

Occupation	Category	Bongas		Sarinagen	
		Before Resettlement	Present	Before Resettlement	Present
Farmer	Self-employed	74 (76.3%)	11 (11.3%)	47 (94%)	14 (28%)
	Share-cropper	7 (7.2%)	43 (44.3%)	2 (4%)	26 (52%)
Employee	Public offices	6 (6.2%)	8 (8.2%)	6 (12%)	2 (4%)
	Private sector	3 (3.1%)	1 (1%)	0	0
Labourer	Farm	5 (5.2%)	4 (4.1%)	0	2 (4%)
	Unskilled manual	4 (4.1%)	1 (1%)	1 (2%)	0
	Construction	1 (1%)	2 (2.1%)	2 (4%)	2 (4%)
Unemployed		1 (1%)	23 (23.7%)	0	7 (14%)
Others		27 (27.8%)	23 (23.7%)	4 (8%)	15 (30%)

Table 2.   Floating cage net (FNC) aquaculture ownership.

Ownership	Bongas	Sarinagen
Households owning FNC		
in 1985–1987	55	18
in 2011	11	4
No. of cages/household:		
in 1985–1987	15	7
in 2011	7	8

engaging in FNC aquaculture in the Sarinagen area are much fewer compared to those in the Bongas area (Table 2), although resettlers in Bongas who reported cage possession were fewer than expected. Interestingly, some interviewed resettlers explained that they have only a small part of the FNC population; it is now mostly owned by non-resettlers who have capital, which generally comes from outside Saguling. The data on FNC ownership in the two villages also indicate inequality of access among the resettlers to the designed cage aquaculture programme.

Convenience of livelihoods

Generally, the resettlers who live either in Bongas or Sarinagen came from the same villages or from neighbouring villages. Geographically, the new sites they have chosen are not far from the previous ones. Therefore, the new settlements do not differ much in access to or networking with public infrastructure. The distance between their dwelling sites to the main road can be considered unchanged; nevertheless, some resettlers in Bongas reported that their dwelling sites are farther from the main road compared to those in their original settlement. This is understandable because their new dwelling sites are isolated from the previous main road by water; overland access is now disconnected. As a result, some important facilities such as administration offices, health units, markets and the downtown are effectively farther away, because they have to use a different route on land. It is possible to follow the old route by using the waterway, but this is more expensive.

A majority of the resettlers from Bongas as well as from Sarinagen reported that their dwelling units are larger compared to those before resettlement. In Bongas, fewer resettlers reported that their dwelling units remained the same or were smaller. In Sarinagen, nearly half of the resettlers reported that the size of their dwelling units had remained the same (Table 3). Immediately after resettlement, they were smaller in size, and mostly non-permanent, with unpaved floors, thatched walls, and tiled roofs; but over these 25 years, the resettlers have constructed better dwelling units, mostly permanent structures characterized by cemented/paved floors, brick walls, and tiled roofs (Table 4). Of the two groups, more resettlers in Sarinagen own non-permanent dwelling units, indicating that more people have a less comfortable livelihood.

Table 3.   Changes in house size in resettlement.

	Change in house size		
	Larger	Same	Smaller
Bongas	61.1%	26.3%	12.6%
Sarinagen	50%	48%	2%

Table 4.   Housing style from before resettlement and at present.

Style of housing		Bongas	Sarinagen
Before resettlement	Permanent	21.6%	12.0%
	Semi-permanent	12.4%	2.0%
	Non-permanent	63.9%	86.0%
	Others	2.1%	0
Present	Permanent	85.6%	80.0%
	Semi-permanent	3.1%	0
	Non-permanent	9.3%	20.0%
	Others	2.1%	0

In their original settlement, the people had no electricity – the situation for most of those in rural areas in Indonesia. Only a few enjoyed electricity using battery accumulators. Since the dam construction and the development of Saguling hydropower, almost all the resettlers have electricity supplied by the PLN. Only very few resettlers in Bongas remain without electricity.

With regard to drinking water, in their original settlement water supply was not available. Even 25 years after project development, they do not have such service yet. The people have been depending on wells, pumps, lakes and rivers, and water springs to fill the drinking-water demand. However, the number of wells and pumps is much larger in the new settlement. Many resettlers own wells or pumps so that they can extract water easily, or they receive a share from neighbouring households. In contrast to the number of wells and pumps, the number of spring-water users has decreased. As development continues in the country, the introduction of water-pump technology into both villages is increasing. Recently, the technologies of bottled water have also been appearing in rural areas such as Bongas and Sarinagen. Some resettlers prefer to consume this water, which is of better quality than well or pump water, even though buying water is more expensive. Drinking-water availability seems to make the resettlers feel more comfortable in their new settlement.

Education

The educational status of the resettlers' children represents the level of education after the resettlement. It was evaluated based on the availability of school facilities and opportunities for higher education and employment. In general, education status can be considered greatly improved: nearly 90% of the resettlers, both in Bongas and in Sarinagen, reported that availability of school facilities for their children was far better than before resettlement. This may correlate to the "obligatory nine-year schooling" programme from the national government, which makes primary school (SD) and junior high school (SMP) compulsory for citizens. Consequently, large-scale construction of such school facilities has taken place in rural areas such as the Saguling area. A majority of the resettlers in Bongas and Sarinagen also perceived that the opportunity for higher education was much improved, as well as the opportunity for better employment. They believed that the formal education programme had improved the education status of their children.

Social life

The social cohesion of a community can be indicated by the time and resources devoted to religious activities. The resettlers both in Bongas and Sarinagen participate actively in building mosques as a common praying facility for most of the resettlers and non-resettlers.

The majority of mosques in Bongas and Sarinagen were built by the local people on a self-supporting basis, while a few were supported by the local government. In Bongas, a majority of resettlers perceive that their religious activities have increased with more participation, while in Sarinagen they reported no change in the level of such activities. It seems that assimilation between resettlers and non-resettlers is better in Bongas than in Sarinagen.

General satisfaction

As indicated above, it was revealed from interviews that the resettlers, both in Bongas and Sarinagen, enjoyed the new environment with increased quality of livelihood. Changes in house size and style, connection to the electric grid, availability of drinking water, health facilities and education facilities were among the measures that had improved compared to their living condition before resettlement. However, the resettlement programmes set up by the PLN and the local government seemed not to be fulfilling their expectations completely. Employment was one of their main concerns; among the resettlers who live in Bongas there was a decrease in satisfaction level with the present jobs, while among those live in Sarinagen there was a slight increase. The data indicate (see Table 1) that many resettlers lost their original job as self-employed farmer and became share-croppers. This might be related to the larger number of unemployed in Bongas as well. The type of employment in which the resettlers participate will affect the level and stability of income, and hence the economic condition. Therefore, a feeling of dissatisfaction with the present job in Bongas may relate to their dissatisfaction with their present economic condition. In contrast to what happened in Bongas, a majority of resettlers in Sarinagen (56%) expressed their satisfaction with the present economic condition.

With regard to the present living environment, a majority of the resettlers living in Bongas (64%) and in Sarinagen (60%) expressed satisfaction. A previous study revealed that the displaced people preferred to live in the vicinity of Saguling Dam rather than accepting the resettlement schemes provided by the PLN and the local government (Gunawan, 1992; Wikarta, 2011). This may limit their stress because they still share the same cultures and customs. However, a majority of the resettlers do not seem to have access to public facilities such as community halls, common toilets, volleyball and badminton courts and football fields. In rural areas, the development of such public facilities seems not to get prioritization from the local government, compared to infrastructure such as roads, bridges, and irrigation facilities.

A majority of the resettlers who live in either Bongas or Sarinagen feel happy with respect to the present environment, which promotes the interests of their children. They perceive that their children have better opportunities to get a good education, so that in the future they will be better off than in the past. It is also surprising that even though land ownership has decreased, they are satisfied with the land availability now. It may be a socio-cultural case; typically people in West Java like to stay close to their relatives rather than being separated. As long as they can mix with their family members, ownership of property becomes a lower priority. For them, job opportunities for their children are satisfactory at the moment. They are optimistic that their children will have better livelihoods compared to their parents. The resettlers hope that with better education, their children will be exposed to employment opportunities in public offices, schools, military service, private companies, and other sectors.

Conclusions and implications for future resettlement policy

The survey revealed that the construction of Saguling Dam has had long-term socio-economic impacts for resettlers. This is true because about 25 years after relocation the resettlers still suffer from unintended occupational changes; many resettlers involuntarily converted their occupations from self-employed farming to share-cropping, which has lower benefits in terms of income-generating opportunities and social status. After losing their fertile farmland, they were not able to purchase land with the same extent in the new settlement with the compensation money they received because land prices increased as a result of the dam project. The other important finding is that more resettlers remain unemployed, which indicates fewer job opportunities since resettlement. Both these negative impacts seem to be common consequences resulting from involuntary resettlement programmes induced by dam constructions. This leads to socio-economic hardship, as pointed out by IUCN and World Bank (1997).

The provision of a new prospect, aquaculture development, was expected to generate employment and create new livelihoods. Fujikura, Nakayama, and Takesada (2009) reported that the resettlers expressed satisfaction with the arrangement. However, doubts remain about whether resettlers were able to reap the maximum benefits of resettlement. In the short term, the possibility of deriving benefits from aquaculture development was perhaps real because for some time from its commencement the population who adopted FNC increased rapidly. According to Pullin (1990), there were about 1083 families engaged in FNC aquaculture. Effendi (1991) reported a different number: 1232 people participating in FNC aquaculture in 1990. Certainly the participation in FNC aquaculture was not necessarily as owner; it could be as FNC aquaculture labourer, fish trader, seed supplier, or feed supplier. The other possible occupations related to this were *bandar* (broker); feed shop owner or labourer; *calo* (middleman); boat owner or operator; driver or assistant driver; net maker; and spicy-fish producer.

Regardless of the number of people that could be absorbed, FNC aquaculture has made a significant contribution in rebuilding the livelihoods of the resettlers. Up to this stage, many researchers have pointed out the superiority of socio-economic status of the resettlers, who were mostly from the village of Bongas. This study shows that several measures such as property, convenience of livelihood and social cohesion were generally better in the village which had better access to aquaculture.

However, the rapid development of FNC aquaculture did not mean that this farming activity was without constraints. Catastrophic fish mortality became one of the major factors affecting the sustainability of the aquaculture. The unfavourable environment has made incidents of fish mortality very frequent. Among the conditions are: (1) seasonal changes which led to the fluctuation of the water level, creating a critical period for raising fish; (2) poor water quality due to heavy pollution, originating mainly from the Citarum River, which contains poisonous chemicals; and (3) fish diseases caused by virus and bacteria. In the past decade, incidents of large fish kill have occurred several times, causing fish productivity to decrease drastically. Wikarta (2011) has evaluated the effects of the decrease in fish productivity on the income of the fish farmers. It was reported that between 1988 and 2012 the income of the fish farmers decreased continually as fish production decreased.

The present study reveals that during the 1990s many resettlers who owned FNC were not able to reconstruct the aquaculture after incidents of fish kill. FNC aquaculture is capital-intensive; the resettlers had to expend about IDR2.4 million (USD1,270) in cages and operation costs (Gunawan, 1992). In this situation, the capital needed to invest in and operate a few fish cages to secure an income sufficient for a single family was beyond the reach of ordinary resettlers and they were obliged to work as employees rather than owners. After the Asian financial crisis in 1997, it was even more difficult for them to reconstruct their FNC aquaculture. Some interviewees stated that non-resettlers and entrepreneurs had been dominating the ownership of FNC in the Bongas area right from the beginning.

It is suggested that this could be the reason why several researchers have raised serious questions regarding the extent to which the full benefits of aquaculture development are enjoyed by the resettlers. The researchers consider that in the long term, aquaculture development has failed to offer alternatives to indemnify resettlers for their lost livelihoods.

However, in general, the livelihoods of the resettlers did improve in their new settlement. They have been enjoying more infrastructure facilities and services which were not available in their original settlement. Nevertheless, some people were not satisfied with the present jobs and economic conditions, which can be considered to be the most important aspect in the human interest.

To summarize the key findings drawn from the Saguling Dam development project and its resettlement programme:

1. In the new settlement, the livelihood patterns of the resettlers are in general improved, nearly three decades after the resettlement programme.
2. However, the dam development project has also caused long-term negative consequences for the resettlers. Conversion to less preferable and beneficial occupations, and unemployment, were among the most common negative impacts.
3. The alternative option of the resettlement scheme, i.e. aquaculture development, was successful in helping the resettlers restore their livelihood only in the short term. In the long term, aquaculture development failed to provide alternative opportunities to rebuild lost livelihoods because of environmental and socio-economic constraints.
4. Differences in access to and opportunities for resources have created differences in quality of livelihood among the resettlers.
5. Jobs availability and economic conditions seem to be the most important factors affecting the satisfaction level of the resettlers.

Key points for recommendation:

1. Policy makers should consider the long-term consequences of dam construction on project-affected people when a resettlement scheme is planned and implemented. Environmental impact assessment will be instrumental in predicting the future situation and condition after dam construction and implementation of the resettlement programme.
2. Selection of resettlement alternative options should consider local backgrounds, such as socio-economic and environmental characteristics, to avoid inequality among the resettlers. Intensive dialogue with local people and scientists will be helpful to describe the local backgrounds. The planned resettlement schemes need not be the same, as long as equality of access and opportunity among the resettlers is secured.
3. The resettlement programme should also consider the local dynamics in order to make appropriate anticipations for the future in case the programme is not successful.
4. Job provision should be given high priority in the resettlement programme because jobs will be of the greatest interest to the resettlers.

Acknowledgements

The research carried out for this study was funded by the Mitsui & Co., Ltd, Environment Fund. This study was also partly supported by KAKENHI (24310189).

References

Bartolome, L. J., de Wit, C., Mander, H., & Nagraj, V. K. (2000). *Displacement, resettlement, rehabilitation, reparation, and development*. Cape Town: World Commission on Dams.

Cernea, M. M. (2003). For a new economics of resettlement: A sociological critique of the compensation principle. *International Social Science Journal*, 55(175), 37–45.

Effendi, P. (1991). Manajemen waduk untuk usaha perikanan [Dam management for fishing business]. Paper presented at *Temu Karya Ilmiah Pengkajian Alih Teknologi Budidaya Ikan Dalam Keramba Mini* [Seminar on transfer of fish culture technology], 4–6 March 1991, Bogor, Indonesia.

Egre, D., & Senecal, P. (2003). Social impact assessments of large dams throughout the world: Lessons learned over two decades. *Impact Assessment and Project Appraisal*, 21(3), 215–224.

Fujikura, R., Nakayama, M., & Takesada, N. (2009). Lessons from resettlement caused by large dam projects: Case studies from Japan, Indonesia and Sri Lanka. *International journal of water resources development*, 25, 407–418.

Gunawan, B. (1992). *Floating net cage culture: A study of people involvement in the fishing system of Saguling Dam, West Java*, MA thesis, Ateneo de Manila University, Manila.

IOE (Institute of Ecology) (1979). *Environmental impact analysis of the Saguling dam: Studies for implementation of mitigation of impact and monitoring. Report to Perusahaan Umum Listrik Negara, Jakarta, Indonesia*. Bandung, Indonesia: IOE, Padjadjaran University.

IOE-UNPAD & ICLARM. (1989). *Development of aquaculture and fisheries activities for resettlement of families from the Saguling and Cirata Reservoirs. Volume 2: Main report*, Institute of Ecology, Bandung, Indonesia, and International Center for Living Aquatic Resources Management, Manila, Philippines.

IUCN & World Bank. (1997). *Large dams: Learning from the past, looking at the future* Workshop proceedings, 11–12 April 1997. Gland, Switzerland: International Union for Conservation of Nature; Washington, DC: World Bank.

Manatunge, J., Contreras-Moreno, N., & Nakayama, M. (2001). Securing ownership in aquaculture development by alternative technology: A case study of the Saguling Reservoir. *International Journal of Water Resources Development*, 17, 611–631.

Nakayama, M. (1998). Post-project review of environmental impact assessment for Saguling Dam for involuntary settlement. *International Journal of Water Resources Development*, 14, 217–229.

PLN (Perusahaan Listrik Negara). (1989). Rencana pengelolaan dan pemantauan lingkungan (RKL & RPL) PLTA Saguling [Planning for environmental management and monitoring of Saguling hydroelectric power]. Jakarta: Author.

Pullin, R. S. V. (1990). Foreword. In B. A. Costa-Pierce & O. Soematwoto (Eds.), *Reservoir fisheries and aquaculture development for resettlement in Indonesia* (pp. vi–vii). Jakarta: Perusahaan Umum Listrik Negara.

Soemarwoto, O. (1990). Introduction. In B. A. Costa-Pierce & O. Soematwoto (Eds.), *Reservoir fisheries and aquaculture development for resettlement in Indonesia* (pp. 1–6). Jakarta: Perusahaan Umum Listrik Negara.

Sutandar, Z., Costa-Pierce, B. A., Iskandar, R., & Hadikusumah, H. (1990). The aquaculture resettlement option in the Saguling Reservoir, Indonesia: Its contribution to an environmentally-oriented hydropower project. In R. Hirano & I. Hanyu (Eds.), *The Second Asian Fisheries Forum* (pp. 253–258). Manila, Philippines: Asian Fisheries Society.

Suwartapradja, O. S., Arifin, T., Kanum, A., Ansor, & Djumari. (1985). *Pemantauan sosial-ekonomi budaya penduduk pindahan dari bawah ke atas genangan PLTA Saguling.* [Monitoring on socio-economic and cultural aspects of the displaced people in Saguling area.] Bandung, Indonesia: Pusat Penelitian Sumber Daya Alam dan Lingkungan, Universitas Padjadjaran.

Wikarta, E. K. (2011). *Dampak pencemaran sumber daya air terhadap aktivitas ekonomi akuakultur di Waduk Saguling DAS Citarum Jawa Barat Indonesia* [The impact of water pollution on aquaculture economic activities at Saguling Reservoir, Citarum Watershed, West Java, Indonesia]. Dissertation, Universitas Padjadjaran, Bandung.

Resettlement and development: a survey of two of Indonesia's Koto Panjang resettlement villages

Syafruddin Karimi[a] and Werry Darta Taifur[b]

[a]Center for Economic Research and Institutional Development, Andalas University, Padang, Indonesia; [b]Andalas University, Padang, Indonesia

Construction of the Koto Panjang Dam was initiated in response to the rapidly increasing demand for electricity in the central region of Sumatra, Indonesia. The process of resettling the villages affected by this construction lasted from 1991 to 2000. The economic factors related to this resettlement programme include monetary compensation, productive capacity, and appropriate distribution of income. Better-off villages (such as those where a rubber plantation was found) received a higher level of compensation and used this compensation to purchase productive assets. Increasing the level of a family's income generates better income distribution and a lower level of poverty, whereas decreasing it creates worse income distribution and a higher level of poverty. The presence of productive capacity is necessary to guarantee the success of an involuntary resettlement programme that attempts to improve the standard of living for displaced peoples.

Introduction

The Koto Panjang dam construction project was initiated in response to the rapidly increasing demand for electricity in the central part of Sumatra, particularly in the provinces of Riau and West Sumatra (JBIC, 2002). The construction effectively inundated all 10 native villages in the region of Koto Panjang, after the resettlement of approximately 5000 households from 8 villages in Riau Province and 2 villages in West Sumatra Province. The process of this resettlement lasted from 1991 to 2000. Construction of the Koto Panjang Dam was more important to trade and business development than to the interests of the local community. Therefore, villagers had to be enticed into the resettlement programme to allow the construction of the dam.

The risk of impoverishment is a reality in any involuntary resettlement programme. Compensation is generally not sufficient to rebuild a living environment. Moreover, the promise of an improvement in living conditions requires a well-planned development programme (Cernea, 1997). The Koto Panjang resettlement programme has been in operation since the first relocation in 1998.

Early post-project evaluation reports indicated that most of the residents of the resettlement villages (68%) were facing a worse living environment than prior to resettlement. Only 18.2% of the residents of the resettlement villages appeared to be enjoying a better living environment (JBIC, 2004). Another earlier study was conducted with 200 randomly selected households from 4 villages (Karimi et al., 2005). These villages included Koto Masjid, Pulau Gadang and Pongkai Baru in Riau Province and Tanjung Pauh in West Sumatra Province. This study found that the living situation had improved for nearly 60% of residents of the Koto Panjang resettlement. The same was reported for 80% of residents in Pulau Gadang, 70% of residents in Koto Masjid, 60% of residents in Tanjung Pauh and 12% of residents in Pongkai Baru. A study based on secondary data from the Central Board of Statistics (BPS, 2006) found that poor families accounted for 24% of the resettled families; this statistic includes 37% of residents in Tanjung Balik and 18% of residents in Tanjung Pauh. The proportion of poor families in West Sumatra Province is about 6% (Karimi, Nakayama, & Takesada, 2009).

The Indonesian government has focused on a number of issues over the past decade including poverty reduction, creation of employment, and economic development. Decentralization of the budget and the promise of autonomy have empowered local governments to keep their focus on seriously lagging villages. Therefore, the Koto Panjang resettlement villages play a role in the country's development strategy. The present study attempts to assess changes in living conditions and to analyze the factors affecting the living environment. This study is based on a survey conducted in December 2010.

Involuntary resettlement and economic development

Involuntary resettlement has the potential to impoverish significant numbers of displaced people (Cernea, 1997). Such resettlement results in loss of productive capacity due to landlessness, joblessness, homelessness, marginalization, food insecurity, loss of access to common property resources, increased morbidity, and community disarticulation (Cernea, 1997, p. 1569).

The impoverishment of displaced persons is typically the main pitfall in development-caused involuntary population resettlement. Cernea (1997) developed his "risks and reconstruction model" to address this pitfall. Such a tool is necessary to fully understand the nuances of losing one's natural, physical, human and social capital as a result of sudden resettlement. Appropriate rehabilitation programmes are required to restore this loss of productive capacity. It is also important to protect and restore displaced people's livelihoods by implementing an equitable resettlement programme (Cernea, 1997).

Resettlement can cause deterioration in production systems, which results in food insecurity, marginalization and loss of income. Such deterioration may be due to loss of productive capacity such as land, common property resources, jobs, societal health and community articulation. All of this can lead to unstable livelihoods, declining standards of conduct and uncertainty of construction and development (Croll, 1999b). Losing one's property (depending on the value of that property) is typically a main indicator of impoverishment in involuntary displacement (Mahapatra, 1999). Expropriation of land removes the first principle upon which people's productive systems, commercial activities, and livelihoods are constructed. This is the predominant method through which displaced persons are decapitalized and pauperized, as they lose both real and manufactured capital (Cernea, 1997). For those ousted from their former situations it is not enough to simply re-establish these resources; rather, the goal should be to re-establish the residents of resettlement villages in such a way that they can experience sustained economic growth in the future (Schuh, 1993).

Successful resettlement that mitigates the risk of impoverishment requires appropriate design and planning (Croll, 1999a). It is necessary to sustain the pro-employment process of development in resettlement villages because short-term employment will not provide long-term stability. Also, a remote resettlement location with scarce resources hinders the rehabilitation of productive activities. A very close link between relocation and development is an important ingredient for maintaining and improving the stability of the community.

There are four aspects of resettlement that must be in place to mitigate impoverishment: adequate policies, adequate legal attention, adequate planning and adequate organization (Cernea, 1997). It is necessary to develop production-based strategies that assist with employment of residents after resettlement. It is also necessary to plan approaches that avoid negative environmental effects for the next generation, including increased population density and decreased availability of natural resources. Thus far, social justice and equity are absent in resettlement programmes. Therefore, every development programme that entails resettlement must take social equity into consideration to prevent impoverishment and to avoid resettlement without rehabilitation.

Involuntary resettlement is characterized by four stages (Scudder, 1997). The first two steps are relocation and adaptation to new situations and occupations. Along with the emotional and financial importance of readjustment, many residents of resettlement villages experience a drastic decrease in their income and their standard of living. Also, many people cease investing once they know they must leave their community, and therefore income and assets at the time of relocation are often lower than expected. The third step is economic and community development and the fourth step is consolidation. In most cases resettlement projects do not achieve the third stage. Therefore, it is necessary to develop a comprehensive resettlement programme that provides "the gains, not just the pains, of development" (Cernea, 1999, p. 4).

The Pareto optimum criterion can provide great insight with regard to compensation as an incentive for resettlement. Under the Pareto optimum it is impossible to improve one person's standard of living without worsening another's. Some critics feel that the Pareto welfare optimum is unable to explain why displaced individuals often become worse off. The calculation of the Pareto optimum does not consider the social discontent and despair felt by displaced peoples due to the construction of dams (Goodland, 1997). However, it is necessary to include both displaced and host populations as beneficiaries of dam development. It is crucial to design a resettlement plan that ensures just and fair compensation so as to encourage voluntary resettlement. Such compensation improves the living standard of settlers and for the right price voluntary resettlement can be very attractive. There is no just and fair compensation that does not increase prosperity; however, asset replacement through compensation will not prevent settlers' becoming worse off in the future. Therefore, investment creation is necessary to improve the living standard of resettled individuals (Pearce, 1999). Thus far, the focus has been on mitigating and compensating; however, this leaves out a "developmental" approach that assesses the needs of project-affected communities despite the regional importance of constructing, for example, a large dam (Cernea, 2003).

Involuntary resettlement dismantles production systems; those resettled often lose their jobs, productive land, and other income-generating assets. Massive displacement worsens living standards and weakens the local and regional economy (Cernea, 1996). However, if crafted the right way a resettlement programme can improve personal well-being and social capacities. At the same time, resettlement can also affect an individual's access to cultural artefacts, physical health, psychological health, and emotions toward housing and environment (Webber & McDonald, 2004).

Methodology

Previous studies have discussed the best-performing and the worst-performing resettlement villages (Karimi et al., 2005). The present study compared the best-performing and worst-performing resettlement villages (from Riau Province and West Sumatra Province, respectively), with two goals in mind: (1) to clarify the present living conditions of resettled residents compared to their living conditions prior to resettlement; and (2) to associate the present living conditions with some cause so that inferences can be made concerning how to improve resettlement planning in the future.

This study selected Koto Masjid (in Riau Province) to represent better-performing villages, and Tanjung Balik (in West Sumatra Province) to represent worse-performing villages. The village selection is mainly based on income improvement after resettlement (Karimi et al., 2005; Karimi et al., 2009). A random-sampling method was used to select 50 resettled households from each village to participate in a survey. All interviews were conducted in December 2010. Because the focus of this study was to evaluate the performance of involuntary residents, only resettled families were interviewed. If by chance a non-resettled household was contacted, the interviewers skipped to the next family until 50 resettled families in each village had been interviewed.

Findings

The Koto Panjang resettlement villages have been in operation for about two decades. The relocation took place between 1991 and 2000. The first relocation was to Pulau Gadang in 1991 and then to Tanjung Balik in 1992. The final relocation to Pulau Gadang was in 1996 and the final relocation to Tanjung Balik was in 2000. Table 1 shows that the majority of the relocations took place between 1992 and 1993. The two villages in West Sumatra to be resettled were Tanjung Balik and Tanjung Pauh. The official term for a village in West Sumatra is still *nagari*, not *desa* like in Riau and other provinces in Indonesia. The Riau authority has transformed many villages by implementing a specific strategy for rural development. Koto Masjid is a division of the village of Pulau Gadang.

Construction of the Koto Panjang Dam was a part of an Indonesian development programme to generate energy from its expansive water resources. The consequences of the dam site included the relocation of nearly 5000 families. The success of the resettlement was due to the prospect of growth in their new villages. The government promised to provide improved community facilities and infrastructure as well as to provide financial compensation and homes for every resettled family. This was enough to successfully convince the natives of Koto Panjang to relocate. Table 2 details a number of items listed in the government's promise

Table 1. Year of relocation (% of resettlers).

	Village		
Year	Koto Masjid	Tanjung Balik	Combined
1991	6	0	3
1992	92	24	58
1993	0	60	30
1994	2	10	6
1996	0	2	1
2000	0	4	2
Total	100	100	100

to these people, and this study has validated the fulfilment of those promises; all of the relocated individuals now enjoy improved community facilities and infrastructure.

Pulau Gadang, the village where the residents of Koto Masjid once lived, was completely submerged at the beginning of the Koto Panjang Reservoir; Tanjung Balik, on the other hand, is not completely submerged. Settlers now living in Koto Masjid are no longer able to visit their old village as a result of this submersion. Settlers native to Tanjung Balik are still able to visit their old villages and still work with the rubber trees that they had to leave behind. The different opportunities offered to the two villages might explain their current performance. Residents of Koto Masjid have no choice with respect to their past; they must develop their destiny in a new home. Their background might no longer hinder their current and future life. In contrast, residents in Tanjung Balik might still have their current and future life tied to the old village in some way. Thus, residents of Tanjung Balik may not fully develop their resources in the new village because they still have opportunities in their old village.

The Koto Panjang resettlement was carried out following the perspectives of Cernea (1999). However, the decision to relocate was not necessarily forced. Table 3 classifies the degree of freedom perceived by residents when the resettlement option was offered. If they decided to relocate on their own they are classified as voluntary residents. If they agreed

Table 2.   Promises made to resettlers.

Koto Masjid			
Items	Promised	Delivered	Wished
Money (IDR)	5–10 million	5–10 million	None
Land for you	0.5 ha	0.5 ha	2 ha
Land for your child	0	0	1 ha
House	6 × 6 m	6 × 6 m	6 × 6 m
Farmland	2 ha	2 ha	3 ha
Well (tube wells)	100 units	100 units	100 units
Mosque	3 units	3 units	4 units
Cemetery	2 units	2 units	2 units
School	2 units	2 units	5 units
Main roads	Paved	Paved	Paved
Village roads	Paved	Not paved	Paved
Electricity grid	Free	Free temporarily	Free
Tanjung Balik			
Items	Promised	Delivered	Wished
Money (IDR)	5–10 million	5–10 million	None
Land for you	50 × 20 m	50 × 20 m	1 ha
Land for your child	0	0	1 ha
House	6 × 6 m	6 × 6 m	6 × 6 m
Farmland	2.4 ha	2.4 ha	3 ha
Well (tube wells)	100 units	100 units	100 units
Mosque	1 unit	1 unit	4 units
Cemetery	1 unit	1 unit	3 units
School	2 units	2 units	5 units
Main roads	Paved	Paved	Paved
Village roads	Paved	Not paved	Paved
Electricity grid	Free	Paid	Free

Table 3. Agreement to accept resettlement (%).

Did you ever agree?	Village		Combined
	Koto Masjid	Tanjung Balik	
Yes	58	16	37
Yes, but reluctantly	26	64	45
Didn't agree	16	20	18
Total	100	100	100

reluctantly or had to relocate, they are classified as mandatory residents. Mandatory residents accounted for 63% of the individuals in Koto Panjang; voluntary residents accounted for 37%.

The original character of the residents differentiates Koto Masjid from Tanjung Pauh. Voluntary residents in Koto Masjid account for 58% of individuals; involuntary residents account for 42%. In contrast, voluntary residents account for 16% of individuals in Tanjung Balik; involuntary residents account for 84%. According to Cernea (1999), the risk of impoverishment is always present in involuntary resettlement. Therefore, the potential risk of impoverishment is more serious in Tanjung Balik than in Koto Masjid.

There are several factors that motivate families to accept a resettlement programme. In the case of Koto Panjang's residents, the most important motivating factors were land ownership, a new home, cash compensation, and a good location. The degree of influence of each motivating factor is the same for Koto Masjid and Tanjung Balik. Other non-physical motivating factors including business, crafts for children and education for children were not that important. The degree of influence of each of these non-physical motivating factors was also the same for Koto Masjid and Tanjung Balik.

The authors find the resettlement process successful. The process entails a planning stage and many details leading up project implementation. The government has successfully motivated the local natives of Koto Panjang to relocate by providing better possibilities for living,

Koto Panjang's resettlement programme used various types of compensation to persuade local communities to relocate. Such compensation included cash, new housing, land, a plantation, education and health facilities, and social infrastructure. Many of the residents disagreed with the compensation, but many did not complain. It seems that the main complaint was that the cash compensation did not satisfy many of the residents; only 14% of the residents openly accepted everything about the resettlement programme. They report not feeling satisfied with the compensation in general but they did nothing to get their complaints heard. Only 19% of residents sent their complaints to the village head and only 12% of residents sent their complaints to government officials. Finally, less than 1% took their case to court. The difficulty of raising official complaints is one of the social problems perceived by residents in the two villages.

Resettlers respond differently to the compensation. In Koto Masjid, 28% of resettlers stated that they were satisfied with the compensation. In contrast, no resettlers in Tanjung Balik were satisfied with the compensation. Furthermore, no resettlers have expressed agreement with the compensation.

Residents often have important reasons to relocate. However, the reasons are not always independent and voluntary. In fact, the resettlement programme was initiated because of dam construction; therefore, it is the government's responsibility to take care of the affected families.

Table 4.   Access to electricity (%).

Period	Village		Combined
	Koto Masjid	Tanjung Balik	
Before resettlement	20	14	17
Soon after Resettlement	100	20	60
At present	100	100	100

Public infrastructure and facilities

Higher-quality public infrastructure and facilities were included in the resettlement programme promised by the government. It is evident from the field study that residents have enjoyed these improvements; most now have access to electricity. Table 4 shows the rate of access to electricity prior to and after resettlement; access to electricity increased from 17% of residents in Koto Masjid to 60% of residents soon after resettlement. All residents now have access to electricity. Access to electricity was much lower in Tanjung Balik prior to resettlement. Soon after resettlement only 20% of residents had access to electricity in Tanjung Balik. Both villages presently have access to electricity.

Table 5 shows that access to clean water has improved since relocation. The current main source of drinking water is spring water. Residents drinking spring water include 64%; well water, 14%; piped water, 13%; bottled water, 8%; and river or reservoir water, 1%. Prior to resettlement, residents drinking river or reservoir water accounted for 78%; spring water, 21%; and well water, only 1%. The use of river or reservoir water fell soon after resettlement and accounted for 15% of residents. Those drinking well water rose to account for 30% of residents and those drinking spring water rose to account for 55% of residents.

Since relocation, access to drinking water has continued to improve for residents of Koto Masjid and Tanjung Balik. Spring water is the most important source of drinking

Table 5.   Resettlers' sources of drinking water.

Type of Water	Village		Combined
	Koto Masjid	Tanjung Balik	
Before resettlement			
Well	0	2	1
River or lake	66	90	78
Spring	34	8	21
Total	100	100	100
Soon after resettlement			
Well	52	8	30
River or lake	6	24	15
Spring	42	68	55
Total	100	100	100
At present			
Water line	0	26	13
Well	28	0	14
River or lake	2	0	1
Buy	16	0	8
Spring	54	74	64
Total	100	100	100

Table 6.   Quality of and access to public health services (%).

Quality	Village		Combined
	Koto Masjid	Tanjung Balik	
Better quality but difficult access	6	0	3
Better quality and easier access	94	98	96
Poor quality but easier access	0	2	1
Total	100	100	100

water in both villages. Spring water supplies 54% of residents in Koto Masjid and 74% in Tanjung Balik. Well water supplies only a minority of residents in Koto Masjid and has been completely replaced by piped water in Tanjung Balik. Koto Masjid has no piped water facility, but residents drink bottled water, thanks to the increased income of the village when compared to other resettlement villages.

Like electricity and drinking water, health services are improving. Table 6 reports residents' perception of the quality of public health services in the Koto Panjang resettlement villages. Health services are presently higher in quality and easier to access. The proportion of residents receiving better health services in terms of both quality and access accounts for 96% of all surveyed residents; this breaks down into 94% of surveyed residents in Koto Masjid and 98% in Tanjung Balik. A small minority agree that they now have better quality in health services, but suggest that access to these health services is not easier. Another minority agree that they now have easier access to health services but suggest that these services are not of higher quality. These groups account for just 4% in total (6% in Koto Masjid and 2% in Tanjung Balik).

In addition to better access to electricity, drinking water, and health services, residents also enjoy the benefit of better schooling for their children. Table 7 reports the residents' perception of the schooling facilities. A majority (84%) of surveyed residents indicate that the schooling facilities for their children are far better than they were prior to resettlement. Just 12% of residents indicate that the schooling is slightly better than before, 3% indicate that it is the same, and 1% indicate that it is poorer than before.

Economic performance

Occupation

Most residents surveyed indicated that they worked at their own farm both prior to and after resettlement. The importance of a self-employed farm is related to the presence of a rubber plantation. It is evident from Table 8 that the number of residents working on their

Table 7.   Quality of children's schooling facilities (%).

Quality	Village		Combined
	Koto Masjid	Tanjung Balik	
Far better	88	80	84
Slightly better	8	16	12
Same	4	2	3
Poor	0	2	1
Total	100	100	100

Table 8.   Occupations of resettlers (%).

Occupation	Village		Combined
	Koto Masjid	Tanjung Balik	
Before resettlement			
Self-employed farmer	80	82	81
Share-cropper	4	10	7
Public employee	6	0	3
Private employee	2	0	1
Others	8	8	8
Total	100	100	100
After resettlement			
Self-employed farmer	90	86	88
Share-cropper	2	6	4
Public employee	2	0	1
Unemployed	0	2	1
Others	6	6	6
Total	100	100	100

own farm has increased, from 81% prior to resettlement to 88% after resettlement. The proportion of residents in Koto Masjid working on their own farm increased from 80% prior to resettlement to 90% after resettlement; that in Tanjung Balik increased from 82% prior to resettlement to 86% after resettlement. Other occupations that resettled residents are engaged in include share-cropping, working in the public sector, and small trading. However, these occupations are becoming insignificant as of late because residents are working more on their own farms.

Sources of income

It is evident from Table 9 that the rubber plantation is becoming quite important in the economy of Koto Panjang. Rubber as the first source of income has contributed to an increase in the standard of living and is restoring the main economic activity of residents. The role of the rubber plantation as the first source of income includes 79% of residents prior to resettlement and 80% after resettlement. In terms of Koto Masjid, this marks an increase from 78% prior to resettlement to 90% after resettlement, while in Tanjung Balik there was a decrease from 80% prior to resettlement to 70% after resettlement. The Koto Panjang economy traditionally depends on the rubber plantation. In line with the development of a dominant economic structure, the Koto Panjang resettlement programme planned a rubber plantation of 2 ha for every resettled family. The project expected to transfer an already productive 2 ha of rubber plantation to every resident at the start of resettlement in new villages. Residents expected to own productive rubber trees in their new village.

However, the above expectation appears to have been empty. The planned rubber plantation did not take place due to poor project monitoring. This disappointed many residents and led to a public outcry to the authorities responsible for the resettlement project. The Indonesian government eventually responded to the disappointment caused by this implementation failure (Karimi et al., 2009). A new rubber plantation was financed by a national budget in both Tanjung Pauh and Tanjung Balik. The present rubber plantation in Tanjung Balik was productive much later than in Koto Masjid. The rubber

Table 9.   Resettlers' first sources of income (%).

| | Village | | |
Source of income	Koto Masjid	Tanjung Balik	Combined
Before resettlement			
Rubber plantation	78	80	79
Public salary	6	0	3
Private business	6	12	9
Wage labour	8	8	8
Others	2	0	1
Total	100	100	100
After resettlement			
Rubber plantation	90	70	80
Fishing	2	2	2
Public salary	6	0	3
Private business	2	18	10
Wage labour	0	8	4
Others	0	2	1
Total	100	100	100

plantation is expected to increase in importance as the primary source of income for residents in Tanjung Balik as it is now for residents in Koto Masjid.

The return of the rubber plantation as the first source of income is a stabilizing feature in the economy of Koto Panjang. Rubber is a major export commodity of Indonesia and the increased trade liberalization will open a larger market for rubber along with higher prices. This is great for the local farmers in the resettlement villages.

Other income-earning activities include trading, wage labour, public employment, and fishing. Among these activities, trading is the most important as a first source of income. Prior to resettlement, trading was an important source of income for 9% of residents. Since resettlement, this has increased to 10%. The role of trading in Koto Masjid fell from 6% of residents prior to resettlement to 2% after resettlement, reflecting the increasingly dominant role the rubber plantation. The role of trading as the first source of income in Tanjung Balik rose from 12% prior to resettlement to 18% after resettlement. The strategic position of Tanjung Balik (on the national highway connecting Padang, the capital of West Sumatra Province, and Pekan Baru, the capital of Riau Province) has been instrumental in expanding trade activities.

Second sources of income

The availability of a second source of income remains important, even if it is decreasing, for residents in the Koto Panjang resettlement economy. Table 10 shows the decreasing availability of a second income source from 62% of residents prior to resettlement to 49% of residents after resettlement. This trend is true for Tanjung Balik, but not for Koto Masjid. The availability of a second income source in Tanjung Balik fell from 74% of residents prior to resettlement to 34% after relocation. The availability of a second income source in Koto Masjid rose from 50% of residents prior to resettlement to 72% after resettlement.

The structure of a second income source suggests that family income for residents in Koto Masjid is more stable than for residents in Tanjung Pauh, because more residents in

Table 10. Resettlers' second sources of income (%).

Source of income	Village		
	Koto Masjid	Tanjung Balik	Combined
Before resettlement			
Not available	50	26	38
Rubber	10	12	11
Fishing	4	2	3
Public salary	4	0	2
Trading	8	20	14
Wage labour	16	4	10
Gambier	0	36	18
Others	8	0	4
Total	100	100	100
After resettlement			
Not available	36	66	51
Rubber	0	4	2
Fishing	36	4	20
Public salary	4	0	2
Trading	12	6	9
Wage labour	0	2	1
Gambier	0	2	1
Food agriculture	6	14	10
Palm oil	2		1
Cattle	4	2	3
Total	100	100	100

Koto Masjid have a second income than in Tanjung Balik. Furthermore, the sources of income are more diversified in Koto Masjid than in Tanjung Balik.

There is a diversification of economic activity that is taking place in both Koto Masjid and Tanjung Balik; this is obvious from the variety of second income sources. These income sources are mostly based on new products since resettlement and are almost completely different from the second-income sources prior to resettlement. This holds true for both Koto Masjid and Tanjung Balik. The role of fishing as a second income source in Koto Masjid increased substantially, from 4% of residents prior to resettlement to 36% after resettlement. Koto Masjid is now famous all over Riau Province for its catfish and smoked fish. Koto Masjid also supplies fish products to other regions. The rise of the fishing industry in Koto Masjid is due to its local water resources and the increasing demand for fish in Pekan Baru and Padang. In addition to fishing, palm oil is providing a second income for residents of Koto Masjid. The dynamic local resettlement economy in Koto Masjid is obvious from the presence of new buildings and new cars that belong to the residents. Local economic performance in Koto Masjid is also attracting many agencies, especially the rich provincial government of Riau and other private companies attempting to benefit from the growing local market.

In contrast to this, Tanjung Balik belongs to the less advantaged provincial government of West Sumatra. The local land is suitable only for food agriculture and gambier. Gambier is the dried sap extracted from leaves and small twigs of the gambier tree and has multiple applications including tanning and dyeing substances, pharmaceuticals, leatherworking, and the textile industry. Gambier is an export commodity of Indonesia and thus very important for the region (Lima Puluh Koto Regency) as a whole, especially the villages of Tanjung Balik and Tanjung Pauh. Foreign investment in gambier processing has made it even more important for the Tanjung Balik economy. Gambier is easy to sell because the processing company accepts gambier leaves directly from local farmers. There are three gambier processing factories in the region, owned by Indian investors. The most accessible factory for residents producing gambier is located in Batu Bersurat, Kampar District, on the banks of the Koto Panjang Reservoir.

Family income level

The level of income earned dictates a resident's capacity to purchase things required to support their daily living. Unfortunately, there is no statistical information on family income in Indonesia. In the case of the Koto Panjang resettlement survey, the question was whether residents are making progress in terms of their individual purchasing power. They were asked about their family income based on the current price level, and asked to judge their income prior to resettlement based on the price index at the time of survey (2010). Basically, the aim was to determine their family income prior to and after resettlement in order to compare the two.

The distribution of family income level is presented in Table 11. Family income is classified from the lowest to the highest, from IDR500,000 or less per month to IDR4,500,000 or more per month. The percentage of residents with a monthly family income below IDR500,000 per month fell from 43% prior to resettlement to 3% after resettlement. The percentage with a monthly family income between IDR500,000 and IDR1,000,000 fell from 39% to 14%. The percentage with a monthly income between IDR1,000,000 and IRD1,500,000 rose from 10% to 13%. The percentage with a monthly income between IDR1,500,000 and IDR3,000,000 rose from 8% to 44%. There were no residents that received a monthly income above IDR3,000,000 prior to resettlement; however, since resettlement this accounts for 26% of households.

A majority of residents have experienced an increase in real income. Prior to resettlement, 82% of residents earned less than IDR1,000,000 per month; after resettlement, 83% of residents earn more than this. This indicates that an improved income distribution within the resettlement economy has accompanied the increase in real family income. In other words, the process of economic growth is functioning and diminishing the impoverishment risk discussed by Cernea (1997).

Table 11.   Resettlers' income levels (%).

Income level (IDR1000/month)	Village		
	Koto Masjid	Tanjung Balik	Combined
Before			
<500	42	44	43
500–1,000	40	38	39
1,000–1,500	8	12	10
1,500–3,000	10	6	8
Total	100	100	100
At present			
<500	4	2	3
500–1,000	4	24	14
1,000–1,500	4	22	13
1,500–3,000	46	42	44
3,000–4,500	16	4	10
>4,500	26	6	16
Total	100	100	100

Koto Masjid and Tanjung Balik are both experiencing an improved income distribution and an increase in real family income. The proportion of residents in Koto Masjid earning less than IDR1,000,000 per month accounted for 82% prior to resettlement. After resettlement, 92% earn more than this. This is also taking place in Tanjung Balik. Prior to resettlement, 82% of residents earned less than IDR1,000,000 per month. After resettlement, 72% earn more than this. Household income levels have increased since resettlement has occurred.

Asset ownership

Improved living conditions have changed the inventory of assets owned by resettled families; prior to resettlement, they owned lower-quality items, whereas after resettlement they own higher-quality items. Table 12 shows an inventory of valuable items possessed by residents prior to and after relocation. Prior to resettlement, many residents owned less expensive items like a radio/cassette player, a bicycle or a black-and-white TV. Very few residents owned more expensive items such as a motorcycle, a motor vehicle, a refrigerator, an attached toilet, an electric fan, or a satellite receiver. After resettlement, the number of families possessing a colour TV, a motorcycle, a satellite receiver, an attached toilet, an electric fan, a DVD/CD player, a refrigerator, a telephone, or gas appliances has increased substantially. Prior to resettlement, no family owned a telephone, a gas appliance or a DVD/CD player. This increase in the value of household assets indicates that there has been an increase in real family income.

Living conditions

Improved living conditions are important in reducing the risk of impoverishment associated with involuntary resettlement. Table 13 shows how residents compare their current living conditions with the living conditions prior to resettlement. Living conditions have improved for more than 70% of residents; specifically, 88% of residents in Koto Masjid and 52% of residents in Tanjung Balik note an improvement in their living conditions. An increased proportion of residents enjoying better living conditions indicates strong economic performance.

Table 12.   Property owned by resettlers (%).

Items	Koto Masjid		Tanjung Balik		Combined	
	Before	Present	Before	Present	Before	Present
TV, color	0	96	0	98	0	97
TV, black and white	34	6	20	0	27	3
Telephone	0	60	0	58	0	59
Radio/cassette	66	50	58	46	62	48
Bicycle	52	40	38	30	45	35
Motorcycle	10	96	16	92	13	94
Motor vehicle	2	14	2	12	2	13
Refrigerator	4	82	2	70	3	76
Attached toilet	20	88	6	94	13	91
Electric fans	8	86	4	76	6	81
Gas appliances	0	54	0	20	0	37
Brass ornaments	0	8	2	12	1	10
Satellite receiver	2	94	0	94	1	94
DVD/CD player	0	84	0	72	0	78

Table 13.   Living condition of resettlers (%).

Is living condition better?	Village		Combined
	Koto Masjid	Tanjung Balik	
Yes	88	52	70
No	12	48	30
Total	100	100	100

Conclusion

The process of economic development following the resettlement programme of the Koto Panjang dam project was stimulated with monetary compensation, increased productive capacity, and an increased level and distribution of income. Better-off villages received a higher level of compensation and used the compensation to purchase more productive assets. More productive capacity (i.e. rubber plantations) was found in more well-off villages. Also, family income is greater and is substantially increasing in better-off villages. Family income in the worse-off villages is slightly increasing in one village but is significantly decreasing in another village. An increase in the level of family income is followed by better income distribution and a lower level of poverty, whereas a decrease is followed by worse income distribution and an increased level of poverty. An improved standard of living is associated with better productive capacity, higher income, better income distribution, and a lower level of poverty. The risk of impoverishment is lower in the better-performing villages, reflecting the better condition of their productive capacity. The presence of productive capacity is necessary to guarantee the success of an involuntary resettlement programme whose goal is to improve the standard of living for displaced peoples.

Acknowledgements

We highly appreciate the support and comments provided by Professor Nakayama at the University of Tokyo, and Professor Fujikura at Hosei University. This study was funded by the Mitsui & Co., Ltd, Environment Fund and supported by KAKENHI (24310189).

References

BPS (Badan Pusat Statistik). (2006). *Selected indicators of socio-economic indonesia*. Jakarta: BPS.

Cernea, M. M. (Ed.). (1996). *Understanding and preventing impoverishment from displacement: Reflections on the state of knowledge*. Oxford: Berghahn Books.

Cernea, M. M. (1997). The risks and reconstruction model for resettling displaced populations. *World Development, 25*, 1569–1587.

Cernea, M. M. (Ed.). (1999). *The economics of involuntary resettlement: questions and challenges*. Washington, DC: World Bank.

Cernea, M. M. (2003). For a new economics of resettlement: A sociological critique of the compensation principle. *International Social Science Journal, 175*, 37–45.

Croll, E. J. (1999a). *Involuntary resettlement in rural China: Field observations*. Washington, DC: World Bank.

Croll, E. J. (1999b). Involuntary resettlement in rural China: The local view. *China Quarterly, 158*, 468–483.

Goodland, R. (1997). Environmental sustainability in the hydro industry. In T. Dorcey, A. Steiner, M. Acreman & B. Orlando (Eds.), *Large dams: Learning from the past, looking at the future*. Gland, Switzerland: World Conservation Union and World Bank Workshop.

JBIC (2002). *Kotapanjang Hydroelectric Power and Associated Transmission Line Project in Republic of Indonesia: Interim report II*. Tokyo: Japan Bank for International Cooperation.

JBIC (2004). *Kotapanjang Hydroelectric Power and Associated Transmission Line Project: Third party ex-post evaluation report*. Tokyo: Japan Bank for International Cooperation.

Karimi, S., Nakayama, M., Fujikura, R., Katsurai, T., Iwata, M., Mori, T., & Mizutani, K. (2005). Post-project review on a resettlement programme of the Kotapanjang Dam Project in Indonesia. *International Journal of Water Resources Development, 21*, 371–384.

Karimi, S., Nakayama, M., & Takesada, N. (2009). Poverty in Koto Panjang resettlement villages of West Sumatra: An analysis using survey data of families receiving cash compensation. *International Journal of Water Resources Development, 25*, 459–466.

Mahapatra, L. K. (1999). Testing the risks and reconstruction model on India's resettlement experiences. In M. Cernea (Ed.), *The economics of involuntary resettlement: Questions and challenges* (pp. 189–230). Washington, DC: World Bank.

Pearce, D. W. (Ed.). (1999). Methodological issues in the economic analysis for involuntary resettlement operations. In M. Cernea (Ed.), *The economics of involuntary resettlement: Questions and challenges* (pp. 50–82). Washington, DC: The World Bank.

Schuh, E. G. (1993). *Involuntary resettlement, human capital, and economic development*. Boulder: Westview Press.

Scudder, T. (1997). Social impacts of large dam projects. In T. Dorcey, A. Steiner, M. Acreman & B. Orlando (Eds.), *Large dams: Learning from the past, looking at the future* (pp. 41–68). Proceedings of World Conservation Union and World Bank Workshop, 11–12 April, Gland, Switzerland.

Webber, M., & McDonald, B. (2004). Involuntary resettlement, production and income: Evidence from Xiaolangdi, PRC. *World Development, 32*(4), 673–690.

A long-term evaluation of families affected by the Bili-Bili Dam development resettlement project in South Sulawesi, Indonesia

Hidemi Yoshida[a], Rampisela Dorotea Agnes[b], Mochtar Solle[b] and Muh. Jayadi[b]

[a]Hosei Graduate School of Social Governance, Tokyo, Japan; [b]Soil Science Department, Faculty of Agriculture, Hasanuddin University, Makassar, Indonesia

A series of surveys and interviews were conducted with families relocated from the site of the Bili-Bili Dam project in South Sulawesi, Indonesia, to remote transmigration areas in the same province. At the time of the survey, all families had received their full amount of cash compensation for relocation. In addition, they had been given an opportunity to join the Transmigration Programme (TP) to receive land and houses for free; however, many suffered from hardships and their strong attachment to their homeland forced them to return. The results of this survey show that families who joined the TP did in fact use their compensation money to purchase small pieces of land and homes close to their original village. Those who were successful and saved money while living in TP areas, as well as those who sold their land in the TP areas, mostly returned to the dam vicinity and were able to purchase land and homes in that area. It is therefore concluded that this resettlement scheme was successful.

Introduction

Most dam construction projects require significant land acquisition and the relocation of families living in the affected area. These dams are often constructed in remote and mountainous regions where the people rely substantially on agriculture as a means to provide for their families. Resettlement is considered an integral aspect of the socio-economic impact of such projects, and thus land-for-land compensation schemes with additional technical assistance for renewing agriculture have often been preferred over cash compensation (World Bank, 2002). In practice, those residing in peri-urban settings or areas with general economic growth may prefer other income-generating options (World Bank, 2004).

This paper discusses the compensation and resettlement scheme of the Bili-Bili multi-purpose dam project on the island of Sulawesi in Indonesia. The Bili-Bili Dam is located on the Jeneberang River, in the district of Gowa, 31 km from the city of Makassar, the capital of South Sulawesi Province. The main purposes of the dam are to control the flooding in Makassar, to supply drinking water to Makassar and its surroundings, and to supply irrigation to the districts of Gowa and Takalar and the city of Makassar. The reservoir's capacity is 270 million cubic metres, which is capable of irrigating over 24,000 hectares of paddy fields and benefitting more than 10,000 farmers. The dam is equipped with a small-scale electric hydropower plant (17 MW) and the reservoir area provides a recreation spot for the community and for tourists visiting a resort in the upriver district.

While the Bili-Bili Dam project spurred economic development in the city of Makassar and the surrounding regions, 2085 families that had lived in the submerged area were displaced. The Ministry of Public Works and the District Government of Gowa reached an agreement with the project-affected families (PAFs) on the issue of land acquisition in 1987; the relocation scheme began in 1990 and was completed in 1997. The scheme was characterized by cash compensation with the option to join the Transmigration Programme (TP), where each PAF was promised two hectares of farmland and a house. All PAFs were free to decide whether to join the TP or to relocate to another location of their own choosing.

In a previous study, the authors conducted a survey of residents of the TP areas and discovered that many residents had quit farming and subsequently returned to the reservoir vicinity of their original homes. They did this despite the improved living conditions of the TP areas when compared to the living conditions of their original homes. The main reason for this return was emotive in nature: they preferred to live in their original home towns (Rampisela, Solle, Said, and Fujikura, 2009). In this study, a survey was conducted of PAFs who had returned from the TP areas to the vicinity of their original homes in the district of Gowa. The goal was to examine their reasons for returning and evaluate their living conditions. The effectiveness of the relocation programme for this particular dam project is also discussed.

Resettlement

The number of PAFs and their destination of relocation is summarized in Table 1. More than half of the PAFs (1079 families, 51.8%) chose to resettle in the vicinity of the reservoir in Gowa district. These particular PAFs were able to purchase land in the vicinity because the amount of compensation they had received was sufficient to buy other land close to their original homes. They enjoyed improved living conditions with respect to housing and electricity and were better off than the original residents (Rampisela et al., 2009). Another 415 families (19.9%) relocated to urban areas. Although it was difficult to identify and thus interview such families, it is likely that they quit farming and changed their occupation, relocating to Makassar or other cities including Sungguminasa.

There were 591 PAFs (28.4%) that chose to join the TP. These PAFs had owned small plots of land or no land at all and did not obtain sufficient compensation money to buy land in the vicinity (Rampisela et al., 2009). Transmigration areas in Mamuju and Luwu were several hundred kilometres away from their original homes; these residents had no option except to join the TP and receive land for living and cultivation (Figure 1).

Table 1.   Relocation destinations of the Bili-Bili Dam's project-affected families (PAFs).

	Destination of resettled PAFs	Number of PAFs	Relocation periods
1	Reservoir vicinity	1079 (51.8%)	1989–1995
2	Urban areas	415 (19.9%)	1989–1995
3	Luwu District (transmigration)	200 (9.6%)	1990–1991
4	Mamuju District (transmigration)	392 (18.8%)	1991–1995
	TOTAL	2,085 (100%)	1991–1995

Source: PPLH Unhas (1998).

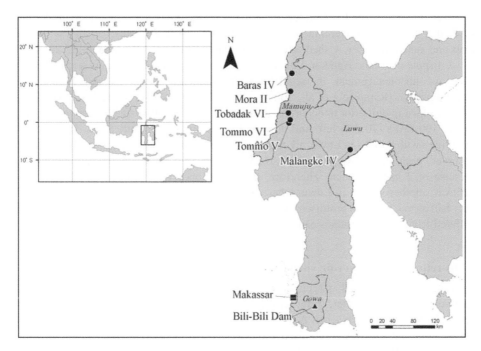

Figure 1.   Location of resettlement areas, the city of Makassar and the Bili-Bili Dam on the island of Sulawesi in Indonesia. *Source*: Prepared by the authors based on International Steering Committee for Global Mapping (2009).

It should be noted that the TP was not developed specifically to accommodate PAFs but rather to accommodate transmigrants from densely populated islands, including Java and Bali. Throughout the twentieth century, the Dutch and subsequently the Indonesian governments set up internal migration policies in Indonesia to tackle the problems caused by population pressure on the land resources in Java and Bali in order that the people could attain self-sufficiency with their rice production. These policies entailed moving populations from the overcrowded islands to the thinly populated neighbouring islands of Sumatra, Kalimantan and Sulawesi. At the beginning of the TP, each transmigrant family was offered a piece of land and tools for slash-and-burn farming (Barral & Poncet, 2007). Presently, every family joining the TP is offered two hectares of farmland and a house; however, they are not allowed to sell the land or the home within the first five years of settlement. This could be an opportunity for landless or poor farmers to become self-employed farmers if they can withstand hardships during the early stages of cultivation.

The Ministry of Transmigration agreed to allocate a certain amount of land for PAFs in the TP areas in the districts of Mamuju and Luwu. The conditions offered to these PAFs were identical to those offered to the transmigrants from Java and Bali.

Some of the infrastructure in Mamuju, including roads and drainage, was not fully developed in the early stages of its receiving transmigrants (around 1991). Prior to the full development of these infrastructures, the PAFs and other transmigrants experienced significant hardship. They often were forced to travel long distances for crop trading; the lack of flood-control infrastructure often resulted in harvest failures. Those who were unable to bear the hardships left Mamuju. Tommo V, the central subdistrict of the Mamuju transmigration area, hosted 194 PAFs (including transmigrants from other islands);

however, 115 PAFs (59.3%) left, leaving only 71 PAFs (40.7%) residing there until 2007. PAFs that were able to overcome the early difficulties were generally satisfied with their lives and stated that they wished to continue living in the area (Yoshida et al., 2010).

A dispute over land rights in Luwu occurred between local inhabitants and PAFs as a result of mismanagement in acquisition of communal lands by the Ministry of Transmigration (Rampisela et al., 2009). Although 200 PAFs registered for the TP, 59 PAFs decided to quit prior to departure for Luwu. Of the 141 PAFs who were relocated to Luwu, 100 PAFs (70.9%) left, and only 41 (29.1%) remained until 2004. According to interviews with 8 PAFs in Luwu, all of them first resettled in a village called Malangke IV; however, each PAF was given only 0.25 hectares of land despite the government's promise of 2 hectares each. Therefore, they were once again forced to relocate – to Sepakat, another village in Luwu, where they were given 1 hectare of land. This is still only half as much as was promised by the TP, but they managed to make it work. All PAFs suffered from water shortages caused by lack of infrastructure.

PAFs returning to reservoir vicinity

A survey was conducted, targeting PAFs returning from transmigration areas to the reservoir vicinity (hereafter, 'returnees'), to examine the role of the TP as an additional option in the relocation scheme. A total of 101 returnees were interviewed at 6 villages in the Manuju and Parangloe subdistricts of Gowa District between December 2010 and January 2011. Returnees in the villages were identified by asking respondents to supply, for the purposes of this study, the names of other returnees they knew; 48 returnees from Luwu and 53 returnees from Mamuju were located. Although no data were available concerning the destinations of PAFs who left the transmigration areas, the number of returnees from Luwu interviewed amounted to nearly half of the PAFs who left Luwu.

The survey covered the following issues:

1. Compensation received
2. Reasons for returning to the dam vicinity
3. Living conditions before, during and after transmigration

The surveyed returnees were either heads of PAF households or members who had experienced life in a transmigration area. Their ages ranged from 24 to 80, with 58% between the ages of 40 and 70. There were 7 returnees who were more than 70 years old. In terms of gender, 67 were male and 34 were female. Most of the returnees had a limited educational background, having completed only primary school or less. The average length of stay in the transmigration area was 3.8 and 2.7 years for returnees from Mamuju and from Luwu, respectively. The combined average was 3.23 years, with 53% returning to Gowa District between 1992 and 1995.

The survey results indicated that 79% of returnees considered their cash compensation in accordance with the promised amount; the remaining 21% considered it less than the promised amount. In terms of overall satisfaction, 70% considered the amount to be satisfactory, while the remaining 30% considered it unsatisfactory. Regarding how they spent their compensation money, 85% of returnees purchased homes and/or land where they presently live. Other uses included food, education and motorcycles. About 10% used the money for pilgrimage to Mecca and 9% spent it on wedding ceremonies.

Most of the returnees from Luwu (40 of 48) stated that their main reason for leaving the transmigration area was land-related problems (Table 2). Twenty-four returnees experienced frequent flooding resulting in a decreased crop yield. Sixteen returnees gave

Table 2.   Families' reasons for returning to the reservoir vicinity.

Reasons	From Mamuju	From Luwu	Total
Low productivity/income/floods	15	24	39
Land disputes	3	16	19
Insurmountable hardship	6	0	6
Desire to reunite with families remaining in the original vicinity	14	4	18
Livelihood unfavorable/safety reasons	3	4	7
Desire for better access to children's education	4	0	4
Health and age-related problems	8	0	8
TOTAL	53	48	101

land disputes as the main reason. Returnees from Mamuju also suffered from floods, but only a small percentage experienced land disputes. The next most important reason for returning was their desire to reunite with wives and children who had accompanied their husbands to the transmigration areas but immediately returned to their original homelands (14 returnees from Mamuju and 4 returnees from Luwu). Eight returnees from Mamuju stated that health and age-related problems were the main reason for returning. Twenty-one returnees from Mamuju were older than 60, while only 7 returnees from Luwu were.

Living situation

Most returnees from Mamuju had owned more land in the transmigration area than they did in their original homelands and present location (Table 3). In contrast, most returnees from Luwu had owned less land in the transmigration areas. The number of those who were landless in their present location increased among returnees from both Mamuju and Luwu.

A common pattern of changes in occupation was observed among returnees (Table 4). In their original homes, many of the returnees were self-employed farmers (SEFs, farmers who work their own lands) and some were tenant farmers. The number of SEFs increased in the transmigration areas and then decreased to the lowest point at their present location.

The above data concerning land ownership and occupation suggest that self-employed farming was not attractive enough to prevent other returnees from leaving the transmigration areas. This analysis is well supported by findings from previous interviews with PAFs in the transmigration areas. Most PAFs remaining in Luwu worked as public officers, including as teachers and policemen. About 40% of PAFs in Mamuju continued

Table 3.   Land ownership of returnees.

	Mamuju			Luwu		
Land ownership (ha)	Original vicinity	TP area	Present location	Original vicinity	TP area	Present location
> 2.0	10	5	4	6	2	4
2.0	2	32	0	0	0	0
1 to <2.0	19	7	6	9	2	3
0.1 to <1.0	17	9	33	32	44	32
0	5	0	10	1	0	9
Total	53	53	53	48	48	48

Table 4. Changes in the occupation of returnees.

	Mamuju			Luwu		
Occupation	Original vicinity	TP area	Present residence	Original vicinity	TP area	Present residence
Self-employed farmer	45	52	35	42	46	29
Tenant farmer	5	1	7	6	0	8
Government employee	0	0	0	0	0	0
Company employee	1	0	0	0	0	3
Trader/businessman	1	0	3	0	0	2
Labourer	0	0	5	0	0	3
Unemployed	0	0	2	0	2	1
Others	1	0	1	0	0	2
Total	53	53	53	48	48	48

living in the transmigration area, having improved their income and living standards by growing cash crops. It is possible that some returnees preferred being tenant farmers rather than SEFs; growing new crops is more risky than cultivating rice under the conventional crop sharing system. Although agriculture is still the dominant livelihood in the reservoir vicinity, 22.9% of returnees worked in non-agricultural sectors as traders, construction labourers, and doing other jobs. This suggests that job opportunities in non-agricultural sectors have increased in the last two decades.

In terms of homes for the returnees, 24% were in very good condition, 68% in moderate condition and 8% in poor condition. When comparing the current conditions to previous ones (both before and during the transmigration period), the current housing conditions are far superior. On the other hand, a minority of returnees (14%) claimed that their pre-transmigration (original) housing condition had been superior to their current homes. A smaller percentage (3%) claimed that the housing conditions during the transmigration period had been superior to their current homes.

A similar tendency was observed regarding utilities and properties/assets (Table 5). Their current conditions were much better compared to the previous conditions; about 52% of the returnees owned a refrigerator, often indicative of upper-class status. This included 31 of the 53 returnees from Mamuju and 21 of the 48 of returnees from Luwu. Owning motorcycles as well as the physical qualities of one's home were often the indicators of upper-middle class or upper-class households (which included 56% of returnees). Even controlling for the contribution of Indonesia's national economic development during the

Table 5. Comparison of assets and utilities of returnees.

	Mamuju			Luwu			Total		
Assets	Original vicinity	TP area	Present residence	Original vicinity	TP area	Present residence	Original vicinity	TP area	Present residence
TV	17	4	51	9	3	44	26	7	95
Motorcycle	9	3	40	5	0	26	14	3	66
Refrigerator	0	1	31	2	0	21	2	1	52
Access to electricity	13	1	52	5	1	41	18	2	93
Own toilet	11	18	47	18	23	42	29	41	89

past decades, it is possible to conclude that the living conditions of PAFs were presently better than before.

Water access, however, had greatly decreased compared to previous conditions. All returnees had had access to wells in the transmigration area, but only 30% had such access at the time of survey; 32% now had access to PDAM (Perusahaan Daerah Air Minum, the regional state water company). A similar tendency was observed when it came to irrigation water access; 46% of returnees claimed that access to irrigation during transmigration was better than their current access. Only 9% claimed to have better irrigation access presently. This water problem was partly attributed to their coming back to the region; at the planning stage of the regional development project, including construction of the Bili-Bili Dam and its irrigation system, there was no consideration of such a large number of PAFs' returning to the vicinity.

For the sake of corroboration and cross-checks, the survey asked questions regarding various levels of satisfaction. The responses were as follows: 80% were satisfied with their current occupation, 92% were satisfied with their current living conditions, and 96% had no plan to relocate. Many of the returnees lived in fairly good conditions, 24% in very good conditions, and 8% in poor conditions. These results were consistent with the level of satisfaction expressed concerning their jobs and living conditions.

Reasons for returning

Among the reasons given for returning to the reservoir vicinity, the most common ones were land disputes and poverty in TP areas. However, their housing conditions and utilities had improved, and many ranked as middle-class citizens in their present villages. Previous to this survey, it was assumed that the returnees were poor, that this was why they had joined the TP, and that they had failed to establish a livelihood in their transmigration area. A deeper analysis of the returnees' behaviour (i.e. their purchasing of homes) revealed what may have been the real reasons for returning. Regarding their present homes, 85% of returnees purchased them between 1990 and 1999. The remainder purchased their houses within two years of returning to the reservoir vicinity (Table 6). About 85% of returnees purchased their present homes using their own funds, while 13% inherited their homes from their parents or received them from the local government (2% moved in with other relatives). This is consistent with their responses concerning spending the compensation money: 64% had purchased land and homes and 21% had purchased only land. Some of

Table 6. Number of project-affected families who were relocated and who purchased their present home, in four time periods.

		Mamuju			Luwu	
Period	Relocation to TP	Returned to reservoir vicinity	Purchased/ built present home	Relocation to TP	Returned to reservoir vicinity	Purchased/ built present home
1 1990–1991			5	48	4	8
2 1991–1995	53	32	18		36	20
3 1995–2000		14	21		6	11
4 2001–2006		7	7		2	9
TOTAL	53	53	51*	48	48	48

*Two of the returnees moved in with relatives and do not own their own homes.

returnees admitted that they had raised additional funds to build their present homes by illegally selling their old homes and land in the TP areas. There were only two returnees who did not own their own homes and they were currently living with relatives.

Based on an analysis of the data, a pattern of behaviour can be identified for returnees from the transmigration areas:

1. They were all relocated to transmigration areas but returned to the dam vicinity because they found the conditions in the transmigration area too difficult and/or they wanted to live closer to their families.

2a. Those who had received enough cash compensation purchased land and/or homes in their original vicinity.

2b. Those who had successfully saved money in the transmigration areas bought land and homes by adding to their compensation cash.

2c. Those who had not been able to make enough money in the transmigration areas returned to the dam vicinity and are currently living with family or other relatives.

3. Much of the land in the transmigration areas was illegally sold to others or given to the recipients' children.

Discussion and conclusion

The resettlement scheme for the Bili-Bili Dam development provided cash compensation for all PAFs and offered the additional option to join the TP. The cash compensation provided many options for the PAFs, and about 70% purchased land close to their original vicinity or relocated to urban areas. Given that the PAFs were not an isolated ethnic minority in the region and also that Indonesia has experienced rapid economic growth throughout the past two decades, even the PAFs who relocated to the reservoir vicinity may still engage in non-agricultural sectors of work.

Cash compensation, however, does not necessarily provide opportunities for the poor. The landless and the small-scale landowners received less compensation and they may not have been able to restore their quality of life. The TP provided such disadvantaged groups with opportunities to gain new land and to increase their income.

While all participants in the TP experienced hardships, 40% of PAFs in Mamuju successfully established their livelihood. Most returnees also obtained land and homes in the reservoir vicinity by taking advantage of the TP and are satisfied with their present conditions. The operation of the TP at Mamuju and Luwu was not satisfactory, and those who participated in the TP suffered from various hardships throughout their first years. However, those who were able to overcome the difficulties and manage to capitalize on opportunities improved their lives to levels that exceeded their pre-transmigrant lives.

The survey found that 8% of returnees lived in poor conditions. This is consistent with findings from other interviews in the reservoir vicinity and in the transmigration area of Luwu. These individuals had been landless in their original location and had a limited capacity to make use of new opportunities because of the low level of human, social and physical capital accessible to them. When faced with land disputes in Luwu, most PAFs returned to the reservoir vicinity; however, those who wanted to return but did not have the money to do it had to remain in Luwu. These individuals are the most vulnerable against major events such as relocations and disputes; they tend to be at the bottom of the social ladder and to have no voice in development programmes. Careful attention should be paid to this group, and necessary assistance, including vocational training and support for children's education, should be combined with cash compensation and other safety-net programmes.

This paper analyzed the life and occupation changes of PAFs from the Bili-Bili Dam project. The resettlement scheme was fairly successful, except for a small number of poor families who benefitted from neither the cash compensation nor the TP. Some families' returning from the TP areas does not necessarily mean that the resettlement scheme was a failure; the return was mostly due to the poor operation of the TP, which caused many problems for the residents and resulted in some of them illegally selling their land. The diversity of the lives of the PAFs mirrored the demographic change currently taking place in Indonesia; Sulawesi has been experiencing rapid economic growth, and the rapid urbanization of the city of Makassar has increased job opportunities in non-agricultural sectors. More people are encouraging their children to obtain higher education and to work in these urban sectors. In this context, cash compensation was relevant to most of the PAFs. The TP gave the disadvantaged group additional support both directly and indirectly, although there was much room for improvement in its implementation.

Acknowledgements

The research was funded by the Mitsui & Co., Ltd, Environment Fund and partly supported by KAKENHI (18310033 and 24310189).

References

Barral, S., & Poncet, E. (2007). *Transmigration policies in South Sumatra, Indonesia, and rural development: Comparative study of two villages.* Paper presented at the Conference on International Agricultural Research for Development organized by University of Kassel and University of Gottingen, Tropentag.

International Steering Committee for Global Mapping (2009). *Global Indonesia.* Retrieved from http://www.iscgm.org/cgi-bin/fswiki/wiki.cgi

PPLH Unhas (Pusat Penelitian Lingkungan Hidup Universitas Hasanuddin) (1998). *Annual environmental monitoring report.* Makassar: Government of the Republic of Indonesia, Ministry of Public Works, Directorate General of Water Resources Development.

Rampisela, A. D., Solle, M., Said, A., & Fujikura, R. (2009). Effects of construction of the Bili-Bili Dam (Indonesia) on living conditions of former residents and their patterns of resettlement and return. *International Journal of Water Resources Development, 25,* 467–477.

Yoshida, H., Shirai, S., Yamazaki, Y., Suda, M., Doi, N., Shimomura, Y., & Fujikura, R. (2010). *Indonesia Bili-Bili Dam Itenjumin no Kurashi ni Kansuru Ichikosatsu* [Living conditions of resettlers from submerged area of Bili-Bili Dam in Indonesia: An analysis with sustainable livelihoods approach]. *Ningen Kankyo Ronshu, 10*(2), 75–90 [in Japanese].

World Bank (2002). *Involuntary resettlement policy, OP/BP 4.12.* Washington, DC: Author.

World Bank (2004). *Involuntary resettlement sourcebook: Planning and implementation in development projects.* Washington, DC: Author.

The livelihood reconstruction of resettlers from the Nam Ngum 1 hydropower project in Laos

Bounsouk Souksavath[a] and Miko Maekawa[b]

[a]Faculty of Engineering, National University of Laos, Vientiane; [b]Wisdom of Water (Suntory) Corporate-Sponsored Research Program, Organization for Interdisciplinary Research Projects, University of Tokyo, Japan

The Nam Ngum 1 hydropower project took place in the early 1970s and displaced about 23 villages and 570 households. This research focuses on two resettlement villages: Pakcheng and Phonhang. A comparison is made concerning the livelihood conditions of these two villages, resettled in 1968 and 1977, respectively. The methodology involved consultation meetings in each village and one-on-one interviews of 100 households (50 households per village). This case study has determined that in terms of family income for these two villages, Pakcheng is significantly more affluent than Phonhang. This is probably because Pakcheng is located along a main road and has far better facilities and irrigation systems.

Introduction

This research looks into the impact of six villages' being resettled and merged into two new villages (Pakcheng and Phonhang) as a result of the construction of the Nam Ngum 1 (NN1) Dam in 1971. The resettlement village of Pakcheng was established in 1968, during the Laotian Civil War, which ended in 1975. After the civil war, the communist Pathet Lao took control of the government, instituting a strict socialist regime. Although relocation to Phonhang was carried out during a time of domestic peace, the overall socio-economic status of Phonhang villagers seems to lag behind that of the Pakcheng population. However, the conditions of a resettlement package should be at the same level for the same dam project. Through a thorough field survey, this research examines the process of resettlements and the critical elements in developing solid new lives for the community members affected by dam construction.

Forty-five and thirty-six years, respectively, have passed since Pakcheng and Phonhang were established. However, there has been no institutional mechanism to monitor the livelihoods and living conditions of these residents. This paper argues that long-term monitoring of resettled villages and measures to accommodate the views of the village residents are necessary (ADB, 1999). In order to address the challenges identified, such as a lack of infrastructure in the form of roads and reliable irrigation systems, collective measures could be taken for negotiations and community development. In fact, the results of this research were used by the villagers, with the authors' assistance, to reach a deal with the local government offices in building a new road and irrigation system for Phonhang, as discussed in this paper. Because this type of household survey for the NN1 Dam had not been conducted before, even by donors such as the World Bank, Japan or the Netherlands, this research will provide valuable data and information to bring to light how the lives of resettlers in these villages were affected by the NN1 Dam.

The Mekong River and the Role of the Hydropower

The Lao People's Democratic Republic (Lao PDR) is a landlocked, mountainous country that shares borders with Cambodia, the People's Republic of China (Yunan Province), Myanmar, Thailand and Vietnam. This area is also commonly referred to as the Greater Mekong Subregion (GMS). Lao PDR is a multi-ethnic society with a level of cultural diversity that is unparalleled elsewhere in the GMS. With a population of 6.4 million (UNDP, 2006), the economy is based primarily upon agriculture, forestry, livestock, fisheries and other natural resources (World Bank, 2006).

Hydropower is a critical resource along the Mekong Basin and has the potential to satisfy growing national and regional electricity needs. The demand for electricity is projected to increase by seven times in Vietnam and to double in Thailand between 2005 and 2020 (MRC, 2009). For some countries in this region, hydropower now represents a major source of export earnings and has the potential for even greater growth in the future. A regional approach to power supply supports the development of hydropower, because it competes effectively with other supply sources in terms of cost, environmental factors and socio-economic aspects. However, only a minor share of the hydropower potential in the Lower Mekong Basin (LMB) has been developed thus far. Thailand has already developed most of its potential on the tributaries, but Laos has developed only a few of its potential projects (MRC, 2002).

As shown in Table 1, hydropower projects with a total installed capacity of 3495 MW are already in operation in the LMB. Projects totalling a further 2774 MW are currently under construction. The potential for 23,492 MW of additional capacity has been identified and lies predominantly in projects within Lao PDR and Cambodia.

Overview of the Nam Ngum 1 (NN1) hydropower project

The Nam Ngum 1 (NN1) Dam, built in 1971, was the third dam constructed in Laos. Two smaller dams (Nam Dong and Selabam) were constructed in 1970. The NN1 dam is the first large dam in Laos and has an installed capacity of 155 MW with a reservoir area of $370 \, km^2$. The NN1 dam was constructed on the Nam Ngum, a tributary of the Mekong River, and is about 100 km to the north of Vientiane by road.

Table 1.   Installed capacity of hydropower projects in the Lower Mekong Basin.

Country	Installed capacity (MW)			
	Existing	Under construction	Planned/ proposed	Total
Laos	1,545	1,758	17,604	20,907
Cambodia	1	–	5,589	5,590
Vietnam	1,204	1,016	299	2,519
Thailand	745	–	–	745
Total	3,495	2,774	23,492	29,761

Source: Based on MRC (2009), updated by the authors.

Plans for the NN1 hydropower project began in 1957 after the formation of the Committee for the Coordination of Investigation of the Lower Mekong Basin. The NN1 fund, which included Japanese funds, was established in 1966 to finance the first stage of the project. The World Bank administered this fund. Stage I construction started in 1968/69. Stage I was commissioned with a full supply level of 202.5 m above sea level in December 1971. Phase II of this project (Construction and Supplement of Stage II Fund) began in January 1976 and was commissioned in October 1978. The full supply level reached 212 m above sea level. The Development and Management of Fisheries in the Nam Ngum 1 Reservoir project was initiated in 1979, and funded by the government of the Netherlands until 1982. This was followed by Stage III, the installation of an additional 40 MW. The entire project became operational in September 1984 (Nippon Koei, 1972).

Approximately 23 villages were settled in the Nam Ngum region. Within these villages, 570 households and 3242 people were affected by the construction of the NN1 Dam, which inundated about 2840 hectares (ha) of land and 1840 ha of paddy fields (Schaap, 1974). As stated in the final report on Nam Ngum Hydro-Electric Project (Nippon Koei Co., Ltd, 1972) there were no social or environmental impact assessments conducted prior to construction of the dam. This may be due to the effect of the Indochina war, the absence of environmental regulation at the time (Lohani et al., 1997), and the fact that people were less concerned with the social and environmental impact given their belief that the dam was important in the development of the country.

The field survey

Objectives

The objectives of this study are to compare and clarify the present livelihood conditions of the NN1 hydropower project resettlement villages with their livelihood conditions prior to resettlement. This research aims to inform processes for improving future resettlement planning.

Methodology

Primary data was collected through literature review, interviews with concerned authorities, project experts and the residents of the NN1 resettlement villages, and observations of the livelihood conditions in the resettlement areas. Focus-group discussions were conducted for those informants. Consultation meetings took place in each village, and one-on-one interviews of 100 households (50 households per village) were conducted. Questionnaire forms were used for interviewing villagers.

Focus villages: Pakcheng and Phonhang

This research focused on the resettlement villages of Pakcheng and Phonhang. These two villages were selected in order to compare the livelihood conditions between two resettlement villages established in different periods, before and after the end of the Laotian Civil War. Fifty households in each village (Pakcheng and Phonhang) were interviewed. The focus-group interviews used for this survey were conducted with first- and second-generation residents of the resettlement villages. First generation refers to family heads and their children that are over or around the age of 50 and who could remember their life in the old villages (43 households in Pakcheng and 44 in Phonhang). Second generation refers to those who were born after the resettlement (7 households in Pakcheng and 6 in Phonhang).

Demographic pictures of Pakcheng and Phonhang

Pakcheng resettlement village was the result of a merger of four old villages: Na Luang, Konsui, Na Khea and Na Leang. Pakcheng was resettled in 1968. Residents were first resettled in the Thalat area for about three years and then moved to the present Pakcheng Village. The resettlement policy was planned and implemented by the government of Lao PDR. The present village profile consists of 161 households; 105 households were resettled from the old villages while the remaining 56 households are second-generation residents. Of the survey and interviews conducted with 50 households (about 24% of all total households), 43 of the households had resettled from old villages. The remaining 7 households are second-generation. The average size of the 50 households interviewed in this village was about 5.5 persons per household.

Phonhang resettlement village was the result of a merger of two different old villages: Kengnoi and Na Luang. (Residents from Na Luang Village were resettled into both Pakcheng and Phonhang.) The distance between the two villages, Pakcheng and Phonhang, is about 1 km. The present village profile of Phonhang consists of 120 households; 100 households were resettled from the old villages while the remaining 20 households are second-generation. Of the survey and interviews conducted with 50 households (about 29% of all total households), 44 of the households had resettled from old villages. The remaining 6 households are second-generation. The average size of the 50 households interviewed in the village was about 5.3 persons per household.

Results of the field survey: the impact of resettlements

Pakcheng, established during the civil war, is better off than Phonhang, set up after the war

In terms of occupation, many residents of the old villages were self-employed farmers who practiced traditional agriculture (e.g. paddy rice field, slash-and-burn agriculture, family garden). There was no private-sector employment in the old villages. Many of the residents of the resettlement villages still work as self–employed farmers; some of the residents, however, do work outside of the village. About 36% of the families in Pakcheng include family members that work outside of the village. This may be a function of the village's closeness to the provincial centre (about 10 km), making it possible for young people to work as public-sector employees, as labourers, and in other service jobs. Most residents of Phonhang are farmers; only about 10% work outside of the village, much less than the residents of Pakcheng (household interviews, June 2010).

A comparison of family income between the old and the new resettlement villages cannot be made since the value of the currency has changed significantly since the resettlement. There were few people living in the rural areas of Lao PDR at the time who were concerned with money because many depended on natural resources. Thus, there were no markets in the villages and it was difficult to access a town due to distance and road conditions. Contrasting this, each of the current resettlement villages has a market within it. In addition, everyone uses money in his or her daily life. Thus, we were able to conduct a survey of family income in the resettlement villages (household interviews, June 2010).

The average family income in Pakcheng was about LAK23,112,000 per year, twice as much as the average family income in Phonhang at about LAK10,380,000 per year. Some residents of Pakcheng work in the private sector and some also work outside of their village. On the other hand, most people at Phonhang are farmers. The survey found that about 10% of residents of Pakcheng had a minimum family income of LAK3,000,000–5,000,000 while about 22% of residents in Phonhang fell into this category. About 18% of residents of Pakcheng fell into the middle-income bracket of LAK16,000,000–20,000,000 per family while only 4% of residents of Phonhang fell into this category.

The local average income levels were higher than the average poverty line for rural villages in Laos. According to the survey conducted in June 2010, the average income of residents in the resettlement villages was approximately USD2889 per household per year in Pakcheng and USD1298 per household per year in Phonhang. The levels of family income of these resettlement villages are higher than that of the Lao PDR rural poverty line, which was set at USD850 per household per year in 2009. There is a stark contrast between the levels of income and overall wealth between Pakcheng and Phonhang, and the authors argue that the conditions of the resettlement package should be at the same level for the same dam project.

Land use and the amount of land received as compensation

Prior to resettlement, the extent of land used by a family and the right to cultivate that land depended on the number of labourers in each family. There was much land available for families to use and most households had a traditional irrigated paddy rice field. Both Pakcheng and Phonhang have continued the tradition of an irrigated paddy rice field because rice is the main food for each of them. The average land use prior to resettlement was about 1.5 ha per household. In the present resettlement villages the land used for paddy rice fields is about 1.3 ha per household in Phonhang and an average of 1 ha per household in Pakcheng.

Both Pakcheng and Phonhang received the same amount of compensation in the form of land, even though Pakcheng was resettled before Phonhang. The only difference between the two was the quality of agricultural land and the sources of irrigation. Pakcheng is located along the Nam Ngum River, whereas Phonhang is located far away from it. The resettlement policy has provided land for paddy cultivation (about 0.5 ha per household and about $900 \, \text{m}^2$ for a home plot). However, this land use is limited when compared to what was available in the old villages.

The irrigation system in Phonhang is not sufficient

The old villages had traditional irrigation systems that were made with natural environmental resources such as rocks and trees. The old villages also had ample water supply for irrigation given the abundant water resources in the surroundings. In the resettlement villages, this kind of natural irrigation was rare due to a lack of water resources in the surroundings. Pakcheng Village, the first to be resettled (in 1968), had the opportunity to benefit from the Pakcheng Agriculture Project from 1980 to 1995. The project provided an ample irrigation system, pumping water from the Nam Ngum River. On the other hand, in Phonhang Village, resettled later (in 1977), a different irrigation system is used, which has an inadequate water supply due to poor pumping (household interviews, 2010).

Housing in the two villages

There are some differences in the housing styles of Pakcheng and Phonhang, mostly due to the different levels of family income. Many of the houses in Phonhang Village are built of wood and bamboo and the roof is often made of grass. Many of the houses in Pakcheng Village are built with brick, cement and other construction materials. When compared to the old villages, residents of Pakcheng expressed that the houses in the resettlement villages are better while most of those in Phonhang village said that the houses in their old villages had been better.

Phonhang faces difficulties in access to roads and to the market

The old villages had main roads or access roads for transportation. These roads served only as rural roads between the villages and could only be accessed by foot. Some villages located along the Nam Ngum River used the river to visit each other. Thus, residents had to walk about 50–60 km from their old villages to the main road in the district centre. The present Pakcheng Village is located along the main road that leads to the district centre, while the present Phonhang Village is located about 1 km away from main road. Residents of Pakcheng Village sell their products to the market in their village. They can either walk or use bicycles and motorcycles to go to the market. In the case of Phonhang Village, residents travel longer to the market in Pakcheng, which is about 1 km away.

Provisions of electricity, water and sanitation

For the past 40 years the old villages had no access to electricity. Thus, villagers did not have electrical appliances or gadgets. The resettlement villages have electricity and thus each household has adapted to this. Pakcheng, however, having been resettled earlier (in 1968), has had access to electricity since 1976, while Phonhang, having been resettled later (in 1977), only gained access to electricity in 1987.

Residents of the old villages would use the Nam Ngum River's streams, springs and wells as a source of drinking water. Residents of the resettlement villages have to buy water for drinking. Water for other uses, such bathing and washing, can be obtained from the Nam Ngum River and wells. The average cost of drinking water is about LAK3000 per household per day (based on 3–4 persons per household). This is fairly expensive for poorer households that rely on a smaller income. The old villages had no toilet systems; residents would deposit their waste in an open area near their house. Animals such as village pigs would often clean it up. The resettlement villages have toilet systems that are water sealed in every household. Some households have toilets within the house while others have a toilet outside the house.

Children have better opportunities for education in the new villages

The resettlement villages have been built with schools. Phonhang has only a primary school, while in Pakcheng there are both a primary school and a high school. Thus, when children in Phonhang finish primary school they have to go to Pakcheng to enrol in middle and high school. Children walk or use a bicycle or motorcycle to attend the school in Pakcheng, about 1 km away. When considering the placement of their present schools many residents are content since there are primary schools in every village. Children in the resettlement villages can anticipate better opportunities for employment now than before since all of them are enrolled in a school or university for the purpose of working or seeking a new job. This is different from the earlier situation when children attended schools for the sole purpose of literacy. Many children had to leave school to work with the family after they finished primary school.

Resettlers are generally satisfied with the places they live

Residents of the resettlement villages were asked if they were satisfied with their jobs. Most residents indicated that they were satisfied with their jobs, especially the residents of Pakcheng (about 62%). Only 48% of the residents of Phonhang indicated that they were satisfied with their jobs. Residents of the resettlement villages were asked if they were satisfied with the place they live. Most residents of the resettlement villages were reluctant to say that their present village is better than the old village. Many did say that their new village is worse. However, most residents indicated that they are satisfied with the place that they presently live, especially the residents of Pakcheng (about 80%). Only 58% of the residents of Phonhang village indicated that they are satisfied with the place that they live. Residents of the resettlement villages were asked if they want to live there for a long time. Residents of both Pakcheng and Phonhang indicated that they would like to live in their villages for a long time since they do not have other options. As a result of better public infrastructure (e.g. electricity, access roads, schools, public health and water supply), most residents believe that the villages where they live are good for their children. Therefore, most residents want to live in their villages for a long time (household interviews, 2010).

Relationships with the administration and the compensations

Residents of the resettlement villages were asked if anyone had explained the details of the resettlement to them. Residents of Pakcheng Village indicated that prior to the resettlement the leader of their village had explained the details on four to five occasions over a period of two years (1967–1968). However, 47% of residents surveyed indicated that nobody had explained the details to them (about 9% of residents indicated that they did not know). Residents of Phonhang Village indicated that they knew about the resettlement. However, some residents received their explanation from the government in 1968 and other residents received their explanation on two to three occasions in 1977 (the year of resettlement).

Most of the residents in both Pakcheng and Phonhang indicated that they had not had any choice in the resettlement because all the plans depended on the resettlement policy of the government. Most residents in both Pakcheng and Phonhang indicated that they had not had a choice in the location of the resettlement, or that they had not known whether they had a choice. About 88% of residents from Phonhang indicated that they did not have choice in the resettlement process while most residents of Pakcheng (about 53%) indicated that they did not know about the details of resettlement prior to the moves. The resettlement policies and plans were developed by the government without any consultation with the people who were to be resettled. A possible reason for this negative response could be that the NN1 was constructed in 1971 with the main phases of pre-construction and construction occurring during the Indochina war. Therefore, the project could not conduct social impact assessment (SIA) or a resettlement action plan (RAP) for the resettlers before the dam construction. All of the resettlers were forced to leave their old villages in a short period of time, with some families evacuating along the river by boat before the closure of the cofferdam.

Residents of the resettlement villages were asked what was the most important reason for resettlement. For most residents of Pakcheng it was the land and locations that were the most important reasons for resettlement; for most residents of Phonhang it was the land and the houses. Thus far the project has provided land, useful resources, and some other infrastructure development as a form of compensation. Residents of the resettlement villages were asked what was promised, what was actually provided, and what was the quantity and quality of compensation. As mentioned above, the project did not provide cash for compensation and it did not build houses for the resettlers. The project simply provided the land and resources for use in agriculture, temple construction, cemeteries and other infrastructure development. Residents of the resettlement villages were asked if they had taken any action to dispel their dissatisfaction with the compensation. Most of the residents in both Pakcheng and Phonhang had not.

Although there has not been a mechanism to address the views of the resettled villagers, such a long-term monitoring system of relocated villages and measures to accommodate the views of the village residents are of critical importance. In order to address the challenges identified, such as a lack of infrastructure or social services, collective measures could be taken for negotiations and community development.

Main findings of the research, and recommendations

Main road construction for Phonhang Village should be pursued

The present Phonhang Village has only one access road and therefore the residents and visitors cannot get out to other roads except through this one road (Figure 1). Therefore, to improve the livelihood conditions of Phonhang residents, the village should have a main road that runs through the village like that of Pakcheng. A new main road can be constructed from the terminal village of Phonthan to Phonhang and then into Pakcheng before arriving at the provincial centre (Figure 1). If such a new road could be built, visitors from the Vientiane capital traveling to the provincial centre of Vientiane could take this route and reduce the travel time.

In fact, the results of this research were used by the villagers with the authors' assistance to reach an agreement with the local government offices in building a new road and

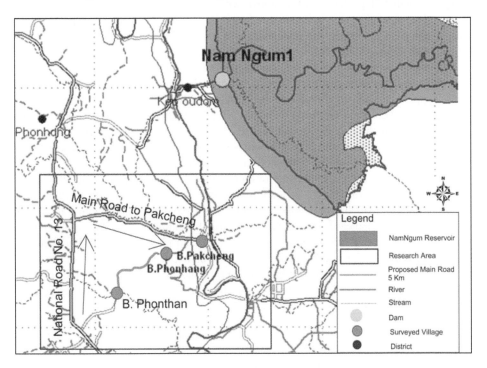

Figure 1. Map of proposed main road through Phonhang Village. Source: prepared by survey team, 2010.

irrigation system for Phonhang. In this negotiation process, there were three key stakeholders from the government side: the head of the Department of Public Work and Transport of Vientiane Province, the Head of the Office of Public Work and Transport, and the head of the Office of Agriculture and Forest of Viengkham District, Vientiane Province.

Improved irrigation system is critical for Phonhang Village

In Phonhang Village, the irrigation system does not have an adequate water supply because of poor pumping and inadequate water resources. Phonhang can only cultivate agriculture in the wet season even though the land is suitable for agriculture all year round. With an improved irrigation system in Phonhang, the villagers could cultivate agriculture in both wet and dry seasons.

The existing irrigation system pumps water from a natural pond and can only supply water to about 10 ha of rice paddy field because the amount of water remaining in the dry season is poor. The entire paddy rice field of Phonhang Village is about 85 ha, much larger than the 10 ha that can be irrigated. Based on the present conditions, Phonhang has two alternatives for developing further irrigation: (1) connecting their irrigation system to the existing irrigation system of Pakcheng, and (2) constructing a new small-scale irrigation system from a second wetland/natural pond.

Current situations of food and income security

The Phonhang resettlement village is located in the Viengkham district of the Vientiane province, which is about 1.5 km from the Nam Ngum River. The current population of this village is about 571 persons (120 families), with an average of 4.8 persons per family. The average population growth rate in this area is currently 2.8%, or about 16 persons per year. If this trend continues the population will exceed 707 persons by the year 2020. All residents of this village belong to the Lowland Lao (*Lao Loum*) ethnic group.

The current occupations of Phonhang residents are as follows: 77% are self–employed farmers; 11% are public-sector employees; 3% are labourers; and the remaining 9% are share-croppers or work in other occupations. Most residents of Phonhang (about 90% of them) work in the agriculture fields within the village, while the other 10% work outside of the village, in the provincial and district centres, in the public or private sector. As mentioned previously, the family income of residents in Pakcheng is greater (by a factor of almost two) thanks to differences in village conditions.

The area of land used in Phonhang for their paddy rice field (approximately 85 ha) is larger than the land available for a paddy rice field in Pakcheng. Residents of Phonhang have an average of 0.7 ha per family (based on 4.8 persons per family with a total 120 families at present). However, residents cannot farm this entire area because of the lack of water from the existing irrigation system. The total land resource used in Phonhang is about 400.5 ha, including paddy rice fields of about 85 ha, vegetable gardens of about 45 ha, fruit orchards of about 64 ha, wetland or natural pond areas of about 19 ha, plantation areas of about 105 ha, grassland of about 67 ha, and other residential land and the village's cemetery of about 15.5 ha.

Proposed irrigation approach development

Based on present conditions, Phonhang Village has two alternatives for developing further irrigation systems. A larger-scale irrigation system should be improved for both rice and other arable crops, because this will be an important contribution that has the potential to improve the productivity of the local farming systems. Compared to the rotation of swidden cultivation, one hectare of irrigated double-cropped paddy can produce up to 60 times more rice than one hectare of swidden cultivation farmed with a 10 rotational bush fallow regime (NAFRI, 2005). The proposed irrigation project would expand these areas according to their command ability by installing permanent connections from the existing irrigation system located in Pakcheng. Also, the project would involve the construction of new diversion weirs from another wetland (Nong Khouay). The existing paddy rice field of 85 ha and the newly irrigated paddy rice field development (approximately 70 ha) would give rise to a total of 155 ha.

Estimated household subsistence needs and output

The average family income per year for residents of Phonhang Village in their present situation is about LAK10,380,000 or about USD1298. This level of total income meets their household subsistence needs only. Thus, residents cannot begin to have any form of savings. Rice is the main food for Lao rural people and an average person needs at least 900 g of paddy rice per day (NAFRI, 2005). Thus far, residents of Phonhang can practise farming but their income meets the household subsistence needs only. Agricultural products (e.g. rice, timber from plantation trees, fuel wood, non-timber forest products, large and small livestock, fish from fish pond and natural pond) are the major output that supports household subsistence needs for residents of Phonhang. Residents also earn some income from the public and private sectors, manual labour, and some other services.

Improvement of family income with a new irrigation system in place

Residents of Phonhang can improve their family income from three main agriculture products: paddy rice, animal husbandry, and vegetable gardens. If the new irrigation system is constructed and water is no longer a limiting factor, residents of Phonhang will be able to farm 155 ha of land (an average of 1.3 ha per household). Under these increased conditions one household could obtain up to 7800 kg of rice products twice a year (in both dry and rainy seasons). However, in addition to the rice products, residents could grow vegetables to increase their family income; the present villages can grow only one crop per year (usually in January). If there is adequate water supply they may be able to grow three crops per year, or at least cultivate $120 \, m^2$ of irrigated paddy rice field per family. Phonhang also has about 67 ha of grassland (an average of 0.64 ha per household). This is suitable for residents to carry out animal husbandry to generate additional income. Table 2 provides estimates for output after additional irrigation development.

Regarding living conditions in the rural areas of Laos, residents must have a family income of at least LAK13,432,900 per year to support their household at an adequate level. If a new irrigation system is installed, the estimated average that one household could save is about LAK9,000,600 from the total output of LAK22,433,500, according to the calculations made by the survey team.

Conclusions

This research suggests that most residents of both Pakcheng and Phonhang are satisfied with life in their present resettlement villages, and they are willing to continue to live there. Many believe that the places they live in are good for their children because of

Table 2. Estimated output after irrigation development.

No.	Description	Area (ha)	Quantity per year	Unit	Unit price (LAK)	Total (LAK)
1	Irrigated rice field	2.6	7,800	kg	1,500	11,700,000
2	Vegetable garden	0.012	720	kg	2,000	1,440,000
3	Large livestock	0.5	292	kg	10,000	2,920,000
4	Small livestock	–	25	kg	20,000	500,000
5	Fruit orchard	0.2	400	kg	2,000	800,000
6	Non-timber forest products	0.5	39	kg	19,000	741,000
7	Fish pond and natural pond	0.01	50	kg	15,000	750,000
8	Building timber		0.5	m^3	1,425,000	712,500
9	Fuel wood		8	m^3	50,000	400,000
10	Public-sector employee					520,000
11	Private-sector employee					430,000
12	Labourer					850,000
13	Other					670,000
	Total				LAK	22,433,500
					USD	2,804

Source: NAFRI (2005).

improved public infrastructure including electricity, road access, schools, public health services and water supply, among others. However, most of the residents felt that they were left with little or no choice concerning resettlement, although residents of Phonhang were resettled during a time of domestic peace and the residents of Pakcheng were resettled during the Laotian Civil War.

Although the NN1 project did not provide cash compensations to the resettlers, it did provide sufficient land and resources. As a result, the resettlers could continue to secure their livelihoods without any cash compensation. Family income for residents of Pakcheng is significantly larger than that in Phonhang. This is probably because Pakcheng is located along a main road and has better village facilities and irrigation systems. Therefore, improving the irrigation systems and constructing a main road through Phonhang is crucial for improving the livelihood conditions for residents of Phonhang.

The authors argue that the conditions of a resettlement package should be at the same level for everyone affected by the same dam project. Therefore, this paper proposes concrete initiatives to rectify the disparities between these two villages by building a new road and irrigation system in Phonhang. This research was actually used by the villagers with the authors' facilitation to agree on new projects with the local government to construct a new road and irrigation system for Phonhang. This paper concludes that long-term monitoring of resettled villages and measures to accommodate the views of the village residents are necessary. In order to address the challenges identified, such as a lack of infrastructure or social services, collective measures could be taken for negotiations and community development.

Acknowledgements

The authors are profoundly grateful to Professor Mikiyasu Nakayama, Professor Masahiko Kunishima, Professor Ryo Fujikura and Dr Hajime Koizumi for their guidance and intellectual support. The authors would like to acknowledge with sincere appreciation that this research was partially funded by the fund of the Mitsui & Co., Ltd, Environmental Fund. This study was also partly supported by KAKENHI (24310189).

References

Asian Development Bank. (1999). *Special evaluation study on the social and environmental impacts of selected hydropower project.* Manila: Author.

Lohani, B., Evans, J. W., Ludwig, H., Everitt, R. R., Carpenter, R. A., & Tu, S. L. (1997). *Environmental impact assessment for developing countries in Asia. Volume 1: Overview.* Manila: Asian Development Bank.

MRC (Mekong River Commission). (2002). *Environmental Training Program (ETP). Revised 2002.* Vientiane: Mekong River Commission Secretariat.

MRC (Mekong River Commission). (2009). *Initiative on sustainable hydropower: Work plan.* Vientiane: Mekong River Commission Secretariat.

NAFRI (National Agriculture and Forestry Research Institute). (2005). *Using of method from agriculture development in Laos.* Vientiane: Author [in Lao].

Nippon Koei Co., Ltd. (1972). *Final report on Nam Ngum Hydro-Electric Project, First Stage. Laotian National Mekong Committee, Mekong Project.* Tokyo: Author.

Schaap, B. (1974). *Report on a household study of Nam Ngum Reservoir evacuees with recommendations for a programme of action.* Bangkok: Mekong Committee.

UNDP (United Nations Development Programme). (2006). *Country report: The Lao PDR.* Vientiane: United Nations Country Office, Lao PDR.

World Bank. (2006). *The Lao PDR environment monitor 2005.* Vientiane: World Bank Country Office, Vientiane & Science Technology and Environment Agency, Lao PDR.

Reconstruction of the livelihood of resettlers from the Nam Theun 2 hydropower project in Laos

Bounsouk Souksavath[a] and Mikiyasu Nakayama[b]

[a]Faculty of Engineering, National University of Laos, Vientiane, Lao PDR; [b]Graduate School of Frontier Sciences, University of Tokyo, Japan

The Nam Theun 2 (NT2) hydropower project displaced 6738 people from 17 villages and 1298 households. This research focuses on four resettlement villages. Household interviews were conducted to learn more about variations in living conditions, traditions and culture in the villages that were relocated independently compared to villages in which relocation had merged older villages together. The case study suggests that most resettlers wanted to remain exclusively with their own village members. However, it was impossible for every village to have its own resettlement location given the scarcity of the land and resources in the resettlement areas. As a result, some villages were merged with other villages in the newly developed resettlement villages. On a different note, the NT2 project provided superior compensation for the resettlers when compared with other similar projects in Laos. However, the NT2 project had insufficient land resources to satisfy the agricultural needs of the resettlers and thus created a situation where the livelihood of the villages will not be sustainable when the project concludes its support for the resettlers.

Introduction

The Lao People's Democratic Republic (Lao PDR) is a landlocked, mountainous country that shares borders with Cambodia, the People's Republic of China (Yunan Province), Myanmar, Thailand, and Vietnam, in an area commonly called the Greater Mekong Subregion (GMS). Lao PDR has a population of 6.4 million. It is a multi-ethnic society, with greater cultural diversity than elsewhere in the GMS. The economy is based primarily upon agriculture, forestry, livestock, fisheries, and other natural resources (UNDP, 2006). Most communities rely on wood for energy and on non-timber forest products for food.

Hydropower in the Mekong River basin has great potential. According to the study results of the Mekong River Commission (MRC), the estimated potential is some 40,000 megawatts. The GMS countries are competing to utilize this potential. More than 200 dams have been proposed for construction on the Mekong River and its tributaries (MRC, 2009). Hydropower development and other water-diversion projects have been initiated in recent years in the Mekong Subregion to generate power, control floods and improve irrigation. More than 60 hydropower projects and more than 70 dams have been proposed in Laos, and so far about 12 dams producing about 1550 MW have been constructed.

The Nam Theun 2 (NT2) hydropower project is built along the Nam Theun River, a tributary of the Mekong River in the central part of Lao PDR. Key features of the project include constructing a dam on the Nam Theun River that will create a 450 km^2 reservoir in the Nakai Plateau, a catchment area of 4013 km^2. The power station will have the capacity to deliver 1070 MW (Nam Theun 2 Project, 2003a), of which 93% is for export to Thailand and 7% for consumption within Laos. The project is forecasted to generate approximately USD1.9 billion in revenue for the government throughout the 25-year project concession period (Nam Theun 2 Project, 2004).

A total of 6738 people in 1298 households were impacted by the project. Among these, some 970 households were fully eligible for the housing and livelihood rehabilitation programme (full impacts), while 94 households were eligible for housing only and 130 other households were eligible for livelihood programmes only (partly affected households) (Nam Theun 2 Project, 2004). Ethnicity was represented by six main ethno-linguistic groups: Brou (40%), Tai Bo (40%), Upland Tai (11%), Vietic (6%), Lao (2%) and Sek (1%). However, distinctions between groups are blurred. Figure 1 indicates the location of the 17 resettlement villages.

During the first decade of the twenty-first century, significant progress was made in project planning modalities, especially with regard to resettlement. The NT2 Resettlement Action Plan (RAP) has been heralded as a model for future large hydropower projects involving resettlements in Laos. Public involvement has evolved more and more with each major development project, and considerable progress has been made, most noticeably in hydropower and road projects.

Objectives

The first objective of this study is to compare the livelihood conditions before and after resettlement of people from four of the old villages: Boua Ma, Done, Sop On and Ca Oy. A second objective is to identify the causes of the problems observed in the present livelihood conditions of the resettlers so that some corrective measures can be planned and implemented.

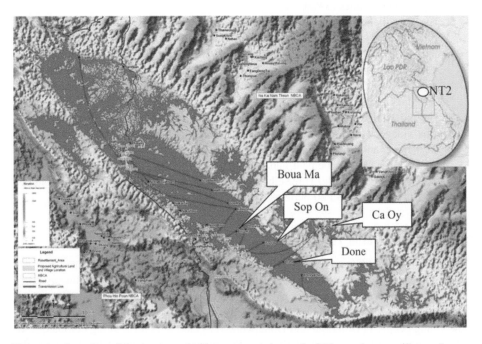

Figure 1.   Location of the 4 surveyed villages among the total of 17 resettlement villages. *Source*: Nam Theun 2 Project (2003b).

Methodology

Primary information was collected through interviews with concerned authorities, interviews and discussions with the resettlers of NT2, and observations of livelihood conditions in the resettlement areas. Focus-group discussions were conducted with three different groups: government agencies, project experts and resettlers. The qualitative information obtained includes discussions with project people and key stakeholders concerning how it has been financially feasible to resettle these people from the reservoir area. The resettlers were interviewed using questionnaire forms.

Focus villages for survey

The NT2 project affected 17 villages. This research conducted household interviews in the four resettlement villages of Boua Ma, Ca Oy, Done and Sop On (Figure 1). The field surveys were conducted from 15 to 25 December 2010 and the sample included 135 households from the 4 villages (50 families who now live in Boua Ma, 35 in Ca Oy, 20 in Done, and 30 in Sop On).

1. The old village of Boua Ma was resettled as Boua Ma (the name was not changed). This village was resettled in 2006. Prior to resettlement, this village consisted of 76 families and 369 people. At the time of resettlement, it received 4 families from other villages, bringing it to 80 families and 386 people. Two ethnic minorities live in this village: there are 31 Brou families (38.75%) and 49 Makong families (61.25%).
2. The old village of Ca Oy consisted of 35 families and 180 people. This village was resettled in 2005. At the time of resettlement, these people were placed into two different resettlement villages: Sop On (31 families) and Done (4 families). The old Ca Oy contained two ethnic minority groups, including 22 Tai Bo families (66%) and 13 Makong families (34%).
3. The old village of Done was resettled as Done (the name was not changed). This village was resettled in 2005. Prior to resettlement, this village consisted of 145 families and 798 people. At the time of resettlement, it received 4 families from the Ca Oy resettlement, bringing it to 149 families and 812 people. Three ethnic minorities live in this village, including 4 Lao families (2.68%), 1 Tai Bo family (0.68%) and 145 Makong families (96.65%).
4. The old village of Sop On was resettled as Sop On (the name was not changed). This village was resettled in 2005. Prior to resettlement, this village consisted of 105 families and 453 people. At the time of resettlement, this village received 31 families from the Ca Oy resettlement and 4 families from other villages, bringing it to 140 families and 633 people. Three ethnic minorities live in this village, including 111 Tai Bo families (79%), 3 Lao families (2%) and 26 Makong families (19%).

Observed changes after resettlement

Occupation and income

Before relocation, the four surveyed villages (in their original locations) were fairly homogeneous in terms of occupations; most people were self-employed farmers, share-croppers or labourers. In the old villages, there were no public or private sectors, although some young people in Boua Ma and Done worked as labourers for logging companies. In the resettlement villages, many of the occupations have been subject to change as the infrastructure and village facilities have changed. Most of the new resettlement villages are located near the main road and so some residents can work with the private and public sectors; however, many remain either self-employed or share-croppers. It must be noted, however, that Ca Oy (before relocation) used to have less access to work outside of the village, unlike Boua Ma before resettlement, which had more opportunities.

Family income

A comparison of average family income in the old villages versus the new resettlement villages is valid since the value of the currency before and after resettlement did not change much. Interestingly enough, the average family income in the old villages was smaller, when compared to the average family income in the resettlement villages. People in the old villages depended heavily on natural resources since there were very few markets and it was difficult to access the town given the distance and the quality of the road. On the other hand, the current resettlement area has a market located within each village.

The surveys conducted in December 2010 have shown that average yearly income is approximately USD1319 per family in Boua Ma, USD1192 in Ca Oy, USD1249 in Done and USD1237 in Sop On. The income associated with all of these resettlement villages is higher than that of the Lao PDR rural poverty line (which was USD850 per household per annum in 2009).

Land ownership and farming activities

In the old villages, traditional irrigated paddy rice fields amounted to about 0.5–1.5 ha per household. Also, the upland rice fields were about 1–2 ha per household. The land use in the old villages depended on the manpower of each household. Households with more manpower used more land than other households with (relatively speaking) less manpower. In the new resettlement villages, however, land availability is limited. The project did allow some land for agriculture, about 0.66 ha per household. The residents are presently using this land to cultivate rice. Thus, they have sufficient rice only for about 2–3 months annually.

Based on the interviews conducted with the villages and the Resettlement Management Unit (RMU), the average amount of land received as compensation is about the same. Home plots are about 800 m^2, while farmland or land for agriculture is about 0.66 ha per household (based on a six-member family). Moreover, the quality of land in the Nakai Plateau, where the old and new villages are located, is comparable. The resettlement villages are located along the present reservoir area of the NT2 project. It has been noted that the quality of land used for agriculture in the Nakai Plateau is not very high when compared to some other zones in Laos.

Animal husbandry

In the old villages, it was a normal occurrence for residents to own buffalos, pigs and cows. These animals were maintained largely as capital assets for use in times of need. Thus, they were not considered a resource that would contribute to regular household income. A small proportion of the adult animals were used to plough in the field in those villages that had permanent rice fields. Buffalos and cows were allowed free-range access to grass fields in the Nakai Plateau (the present reservoir area) and in the forest grazing areas. However, these animals were excluded from grazing in the rice fields and gardens during the periods of cultivation.

The value of one full-grown buffalo was about USD200 before the time of resettlement. Thus, the main income in the old villages came from livestock. Almost every family raised poultry and pigs for household consumption and sometimes for ritual needs. Since the resettlement, the residents continue the practice of animal husbandry. The RMU of the project also found this activity to be the main family income-generating occupation of the resettlement villages. However, it has been acknowledged that the number of livestock is decreasing when compared to the old villages, especially for large animals such as buffalo and cows, as a result of the limited land available.

Irrigation water

There was not a permanent irrigation system in the old villages to support agricultural activities. Therefore, some of the areas could not grow rice even during the rainy or dry seasons. However, there was one weir built near the old villages for gravity drainage; the weir was located at the Nakai Plateau. Unfortunately, the irrigation project was only operational for a few years, after which it broke down (according to residents).

All of the residents stated that since resettlement they now had sufficient irrigation water. However, irrigation water in the resettlement villages of the NT2 project was not intended to supply the lowland rice fields; it was designed for other agriculture cultivation. Given the quality of the land used in the Nakai Plateau, the resettlement villages are not suitable for lowland rice cultivation.

Fishery

Residents of the resettlement villages actively engage in fishing activities. In their old villages, the fish caught in the wet season were larger than those caught in the dry season. However, their quality and weight was better in the dry season. Consequently, residents ate less fish in the wet season.

Fishing in the old villages was carried out in several natural ponds. All of these lakes have a high gradient that results from a series of waterfalls, rapids, riffles and fast runs over stony and rocky substrate, occasionally with sand banks. In the new villages, residents can fish around the reservoir area, which is near the resettlement villages. The reservoir is wider and deeper compared to the lakes before inundation. As shown in Table 1, the average fishing yield increased significantly (from 19 kg to 40 kg annually per capita) after resettlement, thanks to the change of fish productivity, and more fish can be sold to the market.

Table 1. Average fishing yield before and after the resettlement.

Fish species		Before		Present	
Local name	Taxonomic names	Quantity	Where	Quantity	Where
Pa khao	Wallago attu	498	Lakes	1,037	Reservoir
Pa kot	Hemibangrus nemurus	491	Lakes	873	Reservoir
Pa ngone	Laides sp. or spp.	797	Lakes	690	Reservoir
Pa sakang	Puntioplites sp.	327	Lakes	649	Reservoir
Other fish species		1,282	Lakes & ponds	3,365	Reservoir
Fishing yield		2,743 kg/y (19 kg per capita per year)		6,613 kg/y (40 kg per capita per year)	
Fish consumption		1,894 kg/y (12 kg per capita per year)		1,880 kg/y (11 kg per capita per year)	
Fish sold		1,049 kg/y (7 kg per capita per year)		4,733 kg/y (28 kg per capita per year)	

Source: household interviews, December 2010.

Property

Survey results show that household property in the present resettlement villages has increased when compared to the old villages before resettlement. This is due to the availability of better village facilities including electricity, access roads, schools, public health and water supply. Table 2 shows the household property of resettlers before and after resettlement.

Housing

The style of house used after resettlement is different from that used in the old villages. Prior to resettlement, most of the houses were wooden non-permanent or partly wooden semi-permanent. The floor was made of bamboo or timber, thatch and bamboo were used for roofing, and the wall style consisted of timber and bamboo. Homes in the resettlement villages are generally permanent; the floor is made of timber, asbestos is used for the roofing, and the walls consist of timber. In terms of size, some homes in the resettlement villages are larger than those in the old villages, while some are the same size.

Road

The Nakai Plateau, which is the location of the reservoir used for the NT2 project, has access to a major road that runs from the provincial centre to the Nakai District centre. The old villages were located far away from the main road. Boua Ma, for example, was about 12 km from the main road; Ca Oy, about 35 km; Done, about 20 km; and Sop On, 15 km. The resettlement villages are closer to the main road. Boua Ma is now just 3 km from the main road; Ca Oy and Done, about 0.3 km; and Sop On, about 0.2 km.

Facilities

The facilities in the old villages were varied: there was no village market, no clinic, no hospital, only some traditional facilities such as temples and traditional community health centres. In the resettlement villages, residents have assess to a village office and hospital, the community hall, and the village temple. However, some facilities are located in the district centre, including police stations, the local government secretariat, the dispensary and the post office. This was quite far from the four old villages, about 12–35 km; it is also somewhat far from the present resettlement villages, about 10–11 km.

Electricity

Most households in the old villages were without electricity. At that time electricity transmission lines only ran from the provincial centre to the Nakai District. The RMU of the NT2 project provided electricity and other infrastructure facilities to the resettlement villages even before the actual resettlement. Therefore, every household in the resettlement villages has access to electricity.

Drinking water

There were no water-line systems in the old villages. Residents used lakes and streams around their villages for drinking. In the old Boua Ma village, for example, about 26% used a well, 24% used lakes, and 50% used streams. In the old Ca Oy village, 77% used lakes while 23% used streams. In Done, 75% used lakes while 25% used streams. In Sop On, 93% used lakes while 7% used streams. All residents of the resettlement villages use a

Table 2. Household property.

Property	Boua Ma (n = 50)		Ca Oy (n = 35)		Done (n = 20)		Sop On (n = 30)	
	Before	Present	Before	Present	Before	Present	Before	Present
TV (color)	5	38	0	27	2	17	3	22
TV (b&w)	All households use color TV							
Telephone	0	31	0	22	0	18	0	29
Radio/cassette	24	20	11	19	10	12	13	18
Bicycle	21	28	3	13	8	10	7	16
Motorcycle	6	35	0	25	4	12	3	23
Motor vehicle	0	9	0	5	0	4	0	6
Refrigerator	0	39	0	23	0	13	0	1
Attached toilet	0	50	0	35	0	20	0	30
Electric fans	0	41	0	24	0	12	0	18
Gas appliances	All households use fuel wood for cooking							
Brass ornaments	9	21	2	18	6	14	8	13
Satellite receiver	5	38	0	27	2	17	3	22
DVD-CD player	2	31	0	15	1	12	2	14

Source: household interviews, December 2010.

public well that was installed by the NT2 project. There is still, however, no water-line system present.

School

The four old villages did not have properly functioning schools for childhood education. This situation resulted from at least two factors. School attendance was low because the children needed to spend their time gathering forest products to buy rice and other goods rather than attending school. At times, every able-bodied pair of hands was needed to help with the collection of agricultural and non-agricultural produce. Second, teachers did not regularly conduct classes. During the field survey, teachers who should have been conducting classes were often found at home, much to their embarrassment. The incentive for performing one's duty is not large since the district only pays about LAK400,000 per month (in the last seven years before resettlement this was approximately USD30) for rural teachers. The residents themselves financed some schools because the local government had neither the funds nor the resources to provide teachers.

All the residents acknowledged that the educational facilities in the resettlement villages are better than those prior to resettlement. Every village now has a school. Children have more opportunities for higher education. Residents of the resettlement villages expect that their children will have better opportunities for employment after they complete their studies.

Village temple

The residents from the old Boua Ma and Done villages indicated that they each had Buddhist temples. After the resettlement, these two villages still have village temples and the buildings are of comparable size with those in their old villages. The residents of Boua Ma and Done indicated that the village members built the village temples themselves.

All residents indicated that participation in religious activities at the temple was the same in the resettlement villages when compared to the old villages. However, due to differences in belief among ethnic minorities and in the NT2 project resettlement villages, some residents find themselves adhering to Buddhism while others adhere to Animism.

Number of residents

Population growth in the new villages has been on the increase in the last 10 years. The growth rate in the resettled villages currently exceeds 2.5%. As is typical in most underdeveloped areas in Laos, the largest proportion of the population (more than 50% in the NT2 project area) is below the age of 19. This age group is typically the least economically productive, yet it is considered a highly consumptive group. Several factors suggest that there is unlikely to be any significant reduction in population growth in the near future. Some of the notable factors include the lack of birth control use, lack of education about family planning, and the need for large families to provide the labour required for cultivation. Despite the fact that the resettlement action plan of the NT2 project included education on birth control and family planning, the population increase continues, as it does in other parts of the Lao PDR.

Community facility

Regarding the availability of community facilities in the old villages, all the residents answered no, given the location of the villages far from the city and district centre.

For example, the following were lacking in the old villages: commuter buses for going into town, a community health clinic, a police station, a public hospital, a post office, a public library and many other facilities. However, there were some local or traditional facilities that were built and organized by the residents, including temples, village spirits, rural roads between the villages, traditional community health centres and other rural facilities that they could provide for themselves.

The residents indicated that the resettlement villages include many community facilities because these resettlement villages are situated near the district centre. Also, the NT2 project did provide some facilities as a part of the compensation. There are local bus stands/halts, a village administration office, schools, community health clinics, a community hall near the police station, a local government secretariat, a public hospital, a dispensary and other local facilities. The residents indicated that the facilities in the resettlement villages are larger than those in the previous villages (such as the road, school, the village hospital, and others). However, some important facilities (such as the village temple) did not change.

Residents of the resettlement villages were asked if there was common land in their community. The residents answered yes for both before and after the resettlement. They have always had common land or a common forest in their community. For example, in the old village they had community forests, village conservation and protected forests (about 500–1,000 ha per village), as well as a village spirit. In the resettlement villages they also have a similar community forest, about 27,000 ha in total, with 1,000 ha designated for each of the 17 resettlement villages (interview with Mr. Souphaphone, head of the RMU).

Medical and health-related facilities

Residents of the resettlement villages indicated that in their old villages they had not had hospitals or private clinics. There was a small village hospital in the old Done and Boua Ma villages. However, the residents indicated that most of them had used traditional medicine (from forests) and traditional prayer.

The resettlement villages each have a village hospital and some medical workers that were provided by the NT2 project. Therefore, many of the residents use the village's hospital facilities. However, some households still use local medicine (from forests) and traditional prayer. All of the residents indicated that they are more satisfied with the present medical facilities than those available in the old villages. They can now use the village hospital as a public facility as well as the traditional medicine that is available from the surrounding forests. Currently there are no private clinics in any of the resettlement villages.

General satisfaction

Generally residents of the resettlement villages were more satisfied with their jobs in the resettlement community, although some of them responded "do not know" (Table 3).

Most of the residents in these four resettlement villages indicated similar satisfaction with the place they live before and their current village, although some of them responded "do not know" (Table 4). All the residents in the four survey villages indicated that their present economic condition and the current public facilities are better than they had been before the resettlement.

Regarding the resettlement villages, all the residents indicated that the education system and the health facilities were very good for their children. According to the residents, the education and public health facilities in the resettlement areas are much

Table 3. Job satisfaction.

Were/are you satisfied with your job?	Boua Ma (n = 50)		Ca Oy (n = 35)		Done (n = 20)		Sop On (n = 30)	
	Before	Present	Before	Present	Before	Present	Before	Present
Satisfied	35	45	22	31	11	18	23	28
Not satisfied	0	0	0	0	0	0	0	0
Unemployed	0	0	0	0	0	0	0	0
Don't know	15	5	13	4	9	2	7	2

Source: household interviews, December 2010.

Table 4. Place satisfaction.

Are you satisfied with the place you live?	Boua Ma (n = 50)		Ca Oy (n = 35)		Done (n = 20)		Sop On (n = 30)	
	Before	Present	Before	Present	Before	Present	Before	Present
Satisfied	47	48	33	24	15	17	26	28
Not satisfied	0	0	0	0	0	0	0	0
Don't know	3	2	2	11	5	3	4	2

Source: household interviews, December 2010.

better than those prior to the resettlement. However, there were some in each of the four surveyed villages that were concerned about the availability of jobs for their children (Boua Ma 11%, Ca Oy 5%, Done 13%, Sop On 14%).

The residents gave differing expectations regarding the future of their children, as shown in Table 5. Some want their children to continue as farmers, including fishing around the reservoir, forestry and other agricultural activities. Some expect their children to work with a private company, including the NT2 hydropower project or other private companies in the district and provincial centre. Some expect their children to work within the government sector as teachers at schools in their villages, as local government officials, or in the military.

Compensation and relationship with administration

All of the residents indicated that people had indeed explained details of the resettlement to them. However, the residents did not know these people; they were from outside of the

Table 5. Expected job for children.

What kind of job did/do you expect/want your children to do?	Boua Ma (n = 50)		Ca Oy (n = 35)		Done (n = 20)		Sop On (n = 30)	
	Number	%	Number	%	Number	%	Number	%
Farmer	11	22	7	20	3	15	6	20
Company employee	15	30	12	34	8	40	10	33
Teacher	7	14	4	11	2	10	2	7
Public official	13	26	10	29	7	35	11	37
Military service	4	8	2	6	0	0	1	3
Others	0	0	0	0	0	0	0	0

Source: household interviews, December 2010.

village (probably from NGOs or other international organizations). The residents only knew people from the district, and the provincial and central governments that were involved in the NT2 project. The residents received formal information about the resettlement in 1996 as a result of public consultations about the resettlement in 8 of the 17 resettlement villages (Nakai Neua, Nakai Tai, Sop On, Sop Phene, Done, Khone Khen, Thalang and Sailom).

All of the residents indicated that people were interested in negotiation about the resettlement. Also, all of them responded either "myself" or the "leader of village" when naming those who were active in negotiations about the resettlement. However, the practical reality is that most of the residents were not directly involved in the negotiations. The leader officially negotiated for the project.

All the residents indicated that they had been given options for the resettlement. The main options they chose were "provision of land and house" and "near previous residence" because they did not want to live far away from their old villages. (In the four surveyed villages there were no households that chose the "monetary compensation" option.) The reason they chose this option may have been that the NT2 project had a good plan and support for livelihood development in the resettlement areas; they gave guarantees that the livelihood of the resettlers would be better than it once was. Most of the residents indicated that the options did include things that they really wanted. However, some of them answered "do not know".

As shown in Table 6, most of the residents indicated that they had agreed to the resettlement plan – Ca Oy is a notable exception – while some of the residents answered, "yes, but reluctantly". All of the resettlers indicated that the authorities made their promise with a "signed document", without oral or other forms. As shown in Table 7, the promise

Table 6. Resettlement plan.

Did you ever agree to the resettlement plan?	Boua Ma (n = 50)		Ca Oy (n = 35)		Done (n = 20)		Sop On (n = 30)	
	Number	%	Number	%	Number	%	Number	%
Yes	31	62	15	43	15	75	22	73
Yes, but reluctantly	19	38	20	57	5	25	8	27
Did not agree	0	0	0	0	0	0	0	0

Source: household interviews, December 2010.

Table 7. Reasons for moving.

What was the most important reason?	Boua Ma (n = 50)		Ca Oy (n = 35)		Done (n = 20)		Sop On (n = 30)	
	Number	%	Number	%	Number	%	Number	%
Money	2	5	3	11	0	0	0	0
Land	11	28	9	33	5	25	4	13
Place	0	0	0	0	0	0	0	0
House	26	67	15	56	11	55	12	40
Job	0	0	0	0	4	20	14	47
Jobs for children	0	0	0	0	0	0	0	0
Education for children	0	0	0	0	0	0	0	0
Other	0	0	0	0	0	0	0	0

Source: household interviews, December 2010.

of a new house was the most important reason that many resettlers agreed to move, while land use was the second most important reason (a notable exception is Sop On, where jobs was the most important reason). All the residents indicated that when they agreed to the resettlement plan or option they did in fact think about their children. Most of the residents in Boua Ma and Ca Oy considered land use the main concern/interest for their children, while jobs was the second (Table 8). The other two villages (Done and Sop On) considered jobs as the main concern/interest for their children, while land use was the second interest.

Compensation package

Money

The compensation policy of the NT2 project did not involve cash as a form of compensation. Instead, the project provided livelihood facilities and other infrastructure development as compensation (e.g. housing, water supply, agriculture, public health, a village school, a village hall, village roads, proximity to the main road). The project did allow for some cash compensation where there would be losses in agricultural produce (such as fruit trees).

Land for future generations

The project provided land for agricultural use in the resettlement area (about 0.66 ha per household, based on five people per household). The resettlers could only get 0.66 ha as compensation due to the limited availability of land in the resettlement area. In this regard, the land did not meet the wishes of the resettlers. However, the project provided the community with a forest of about 18,206 ha. Each of the 17 resettlement villages received an average of 1071 ha per village. The residents can use this community forest to repair their houses, to gather non-timber forest produces and for other daily uses. They can also sell the timber to generate family income.

House

The project built one house for each affected household of nine or fewer members. In cases where a family had more than nine people they were eligible to receive two new homes. The project built new homes for all affected households. This satisfied the resettlers' wishes because the new homes in the resettlement villages are of superior quality to those in the old villages.

Table 8. Children and resettlement.

What was the main concern/ interest for your children?	Boua Ma (n = 50)		Ca Oy (n = 35)		Done (n = 20)		Sop On (n = 30)	
	Number	%	Number	%	Number	%	Number	%
Land	22	44	14	47	3	15	9	30
Job	14	28	11	37	12	60	12	40
Education	9	18	3	10	1	5	4	13
Health	3	6	1	3	3	15	2	7
Safety	2	4	1	3	1	5	3	10
Other	0	0	0	0	0	0	0	0

Source: household interviews, December 2010.

Public facilities

The project built several wells at each main point throughout the resettlement villages. This satisfied the resettlers' request because the well is close to the houses (a two-to-three-minute walk). The quality of water is also good because the depth of each well is between 25 and 30 m. The project also promised to provide land for a cemetery, a school, village roads and main roads, an electricity grid, wells and free water supply in each village as requested by the resettlers.

Complaints regarding compensation

As described above, the majority of the compensation provided by the NT2 project was not given directly in cash but rather through facility development. However, there was some cash compensation for the loss of tree plantations and other agricultural products. The project did this to reduce dissatisfaction among resettlers with the compensation. Most complaints were made to the village head, while others did not do anything, as shown Table 9.

The resettlement scheme

Most residents indicated that they were satisfied with the resettlement scheme since the infrastructure and other facilities in the resettlement villages are better when compared to those in their old villages. However, it was noted that the residents were not so satisfied with the land available for agriculture in the resettlement villages when compared to their old villages.

Culture and traditional relationships between the ethnic minorities

This section of the interview dealt with the cultural and traditional relationship among ethnic minorities and concerns, mainly with respect to the old village of Ca Oy because this village was resettled into two other villages: Sop On and Done. The old Ca Oy included 35 families and 180 people. Of these, 31 families resettled in the new Sop On while 4 families resettled in Done. Therefore, the following findings concern the three villages of Ca Oy, Sop On and Done most significantly.

Done and Sop On were both merged with Ca Oy during the resettlement. The villagers generally recognized that the merger was planned by the project. A majority of the villagers (80% and 77% in Done and Sop On, respectively) did not mind being merged with Ca Oy villagers. However, when they were asked whether they wanted to be resettled only with those from Done or Sop On, the majority of them (80% and 90% of Done and

Table 9. Complaints regarding compensation.

Did you take any action in order to dispel your dissatisfaction with the compensation?	Boua Ma (n = 50)		Ca Oy (n = 35)		Done (n = 20)		Sop On (n = 30)	
	Number	%	Number	%	Number	%	Number	%
Complained to the village head	38	75	27	77	15	75	24	80
Complained to the government official	0	0	0	0	0	0	0	0
Appealed to the court	0	0	0	0	0	0	0	0
Didn't do anything	12	24	8	28	5	25	6	20

Source: household interviews, December 2010.

Sop On, respectively) answered yes. While about half of the villagers (50% and 54% of Done and Sop On, respectively) felt that they experienced difficulties with the people from Ca Oy, 25% and 33% of Done and Sop On, respectively, did not feel any difficulties.

On the other hand, only 49% of Ca Oy villagers accepted the merger with other villages after relocation, and 37% did not. Among these villagers, 69% wanted to be resettled only with other Ca Oy villagers. While 66% of the villagers recognized that the merger was planned by the project, the rest of them did not. About one-third (34%) of the villagers felt that they experienced difficulties with the people from Done or Sop On, but another one-third (32%) did not. None of the Ca Oy villagers indicated that after resettlement (within 4–5 years) their family members had married anyone belonging to another ethnic group.

Considerations for resettlement development

How can the livelihood condition of the resettlement villages be sustained after the NT2 project suspends its support to the resettlers? The resettlers have decided to live around the reservoir areas and they all will continue to live in these resettlement areas despite the fact that the sustainability of their livelihoods is limited. There are two possible alternatives for improving the resettlement plan: (1) an extension of the land used for agriculture; and (2) the development of a reservoir for aquaculture through an alternative technology.

An extension of land used for agriculture

Table 10 presents land use and forest types around the resettlement area. About 18,206 ha (87.51%) is forested. During the main field survey conducted between March and May 2011, the surveys discussed the possibility of further extension in the land used for agriculture with both residents and authorities of the Resettlement Management Unit (RMU) of the NT2 project. However, to make this possible the RMU must discuss further with the concerned provincial and district authorities in collaboration with the NT2 project before a final decision is made.

The community forest area is sufficient to rezone some areas to serve as agricultural land for the resettlers. The NT2 project has offered land for agricultural use at an average of 0.66 ha per household. In order to produce sufficient rice for year-round consumption the resettlers requested further land for shifting cultivation (about 1 ha per household or about 1298 ha for all resettlement households), in combination with the present land area of 0.66 ha, for a total of 1.66 ha per household (based on six persons per household).

Table 10. Land and forest types in the resettlement area.

Land use/forest type	Area (ha)	Percentage (%)
Forested land	18,206	87.51
Unstocked forest	2,014	9.73
Slash-and-burn agriculture	136	0.66
Swamp	13	0.06
Water body	21	0.10
Residential area	356	1.72
Other	46	0.22
Total	**20,692**	**100**

Source: Household interviews, heads of villages and deputy manager of NT2 resettlement office (Mr Sisaveuy Chanthavisak), December 2010.

With an agricultural area of 1.66 ha, the resettlers could obtain rice production of at least 2480 kg based on a productivity of 1500 kg/ha for rained-paddy fields of one crop per year. To ensure the success of the rezoning and resource use for resettlers and to improve the livelihood conditions of the resettlers, the RMU of the project (in collaboration with resettlers) must discuss the following land and resource use management issues:

1. Consultative village land-use planning
2. Support for approach consultative official land allocation
3. Rain-fed paddy and arable fields
4. Non-timber forest products
5. Wildlife conservation
6. Forest protection

The further request of land for agricultural use is very necessary. With only 0.66 ha/ household, which is what the project provided as compensation, the resettlers cannot obtain sufficient rice for a year. Therefore, the project must consider providing more land for rice cultivation (at least 1 ha per household). This is entirely possible since more than 18,000 ha of community forest is present in the resettlement area.

Conclusions and discussion

Most resettlers surveyed by this project report that they are satisfied with life in their present resettlement villages and that they will continue to live there. Most residents believe that the place they now live in is good for their children because of the improved public infrastructure (e.g. electricity, road access, schools, public health services, water supply). We can also conclude that the family income of resettlers increased when compared to that of the former villages. The annual family income of resettlers was about USD1319 in Boua Ma, USD1192 in Ca Oy, USD1249 in Done and USD1237 in Sop On. Consequently, the family income of resettlers is higher than the poverty line.

Most resettlers sought to be resettled with their old village members. However, this was impossible for every village due to the limitation of land and resources in the resettlement areas. Resettlers also did not want to move far away from their former villages. As a result, some villages had to be resettled and these villages were merged with other villages after relocation. For example, Ca Oy had to be resettled and merged with Sop On and Done because Ca Oy Village was quite small and was located far away from the other villages.

Resettlers were aware of only the short-term positive impact of their resettlement. That is, many of the residents were concerned about housing, because this was of immediate importance for their livelihood. Some residents had hoped to work for the project during the construction period. Only some of the residents were concerned about land and resource use, which was very important in making this resettlement a long-term success.

Extension of the land dedicated for agricultural use is necessary for the sustainable livelihood of the resettlers after project support is suspended in 2014.

The difficulties observed in livelihood rehabilitation of the resettlers largely stem from the fact that resettlers were obliged to convert themselves from a mixture of nomadic and slash-and-burn farming into intensive agriculture with small farmlands. Moreover, water is not constantly available for intensive rice cropping in the newly developed farmlands for resettlers. Their production base was thus largely degraded after relocation. An officer of the Nam Theun 2 hydroelectric project responsible for the resettlement scheme admitted that the existing production base was not sufficient for the resettlers to maintain their livelihood and that some corrective measures are needed, such as giving the resettlers permission to use the public forests for production purposes (personal communication, 13 February 2012).

The authors regard the concession period of 31 years (Nam Theun 2 Project, 2005), of which the operating period is 25 years, as the major source of the observed problems. All the systems needed to be built in 6 years to make the hydropower station operational for 25 years. The area for resettlement was thus confined to a small, newly developed area on the Nakai Plateau (on the shore of the reservoir) that was not appropriate for nomadiss or slash-and-burn farming. Finding a larger area, to have the resettlers maintain their traditional occupation of nomadiss and slash-and-burn farming, was not considered in the development the resettlement scheme because of the paucity of time for planning.

Measures were apparently needed for resettlers to change their jobs, while the productivity of the land given to them was simply not high enough for livelihood rehabilitation. Alternative compensation packages not based on the productivity of farmlands (Fujikura & Nakayama, 2013) should also be developed and offered to the resettlers for their selection.

Acknowledgements

The authors are profoundly grateful to Professor Masahiko Kunishima, Professor Ryo Fujikura and Dr Hajime Koizumi for their guidance and intellectual support. The authors would like to acknowledge with sincere appreciation that this research was partially funded by the fund of the Mitsui & Co., Ltd, Environmental Fund. This study was also partly supported by KAKENHI (24310189).

References

Fujikura, R., & Nakayama, M. (2013). Long-term impacts of resettlement programs by dam construction projects in Indonesia, Japan, Laos, Sri Lanka, and Turkey: Comparison of land-for-land and cash compensation schemes. *International Journal of Water Resources Development, 29*.

MRC. (2009). *Initiative on sustainable hydropower*. Vientiane: Mekong River Commission.

Nam Theun 2 Project. (2003a). *Environmental impact assessment and management plan. Volume 2: Main text*. Vientiane: Nam Theun 2 Hydroelectric Power Project.

Nam Theun 2 Project. (2003b). *Resettlement action plan*. Vientiane: Nam Theun 2 Hydroelectric Power Project.

Nam Theun 2 Project. (2004). *Summary environmental and social impact assessment*. Vientiane: Nam Theun 2 Hydroelectric Power Project.

Nam Theun 2 Project. (2005). *Nam Theun 2 hydroelectric project: Summary of the concession agreement*. Vientiane: Nam Theun 2 Hydroelectric Power Project.

UNDP. (2006). *Country report: The Lao PDR*. Vientiane: United Nations Development Programme.

World Bank. (2006). *Lao PDR environment monitor*. Vientiane: World Bank Country Office.

Long-term perceptions of project-affected persons: a case study of the Kotmale Dam in Sri Lanka

Jagath Manatunge[a] and Naruhiko Takesada[b]

[a]Department of Civil Engineering, University of Moratuwa, Sri Lanka; [b]Faculty of Humanity and Environment, Hosei University, Tokyo, Japan

Many of the negative consequences of dam-related involuntary displacement of affected communities can be overcome by careful planning and by providing resettlers with adequate compensation. In this paper the resettlement scheme of the Kotmale Dam in Sri Lanka is revisited, focusing on resettlers' positive perceptions. Displaced communities expressed satisfaction when income levels and stability were higher in addition to their having access to land ownership titles, good irrigation infrastructure, water, and more opportunities for their children. However, harsh climate conditions, increased incidence of diseases and human–wildlife conflicts caused much discomfort among resettlers. Diversification away from paddy farming to other agricultural activities and providing legal land titles would have allowed them to gain more from resettlement compensation.

Introduction

Over the decades, there has been growing concern about the negative consequences of the involuntary displacement of rural communities for large-scale infrastructure development (De Wet, 2006; Robinson, 2003). The construction of dams is the most often cited example of development projects that cause forced displacement of communities (McCully, 2001). In the short run, the dispossession and displacement of people from their assets, resources, established livelihoods, incomes and social relationships present complex risks and could potentially lead to impoverishment, both socially and economically (Cernea, 1995, 2000; Manatunge et al., 2001; Miyata & Manatunge, 2004). Many studies, largely socio-anthropological investigations, have demonstrated the immediate effects of the displacement caused by development projects (Cernea, 2000; Eriksen, 1999; Pearse, 1999; World Bank, 2001). On the basis of such studies, dams and other large-scale development projects that require the relocation of people have been deemed undesirable investments (McCully, 2001; Scudder, n.d.; World Commission on Dams, 2000). Comparatively little research has been conducted on the longer-term impacts (say, 25–30 years after resettlement) of relocating communities in new environments with convincingly better socio-economic and physical facilities (Mejia, 2000; Partridge, 1989). This is an area that merits careful examination because such research would provide us with information on how successful a community has been in recreating social links with external support and/or its own will, and then rebuilding livelihoods, so as to overcome the difficulties caused by involuntary settlement. Focusing on apparently successful cases would provide guidance on attributes that should be incorporated into future resettlement planning. Conversely, unsuccessful cases would suggest those elements which do not make the resettlers better off after resettlement. However, conditions may vary from place to place, and therefore decisions have to be taken with particular attention paid to the dynamics of the local socio-economic and political systems. A hypothesis worth testing is that in successful cases, the negative short-term consequences of involuntary displacement are offset by longer-term benefits generated from enhanced socio-economic opportunities created in the newly developed site.

The small-holder irrigated settlement schemes in Sri Lanka provide an ideal setting for testing the foregoing hypothesis. As part of the government's agricultural and rural development polices, approximately 150,000 families have been moved, (1) from overcrowded wet zones and (2) from areas affected by dam construction, to irrigated resettlements in the dry zone. These settlement schemes are now well-established farming communities. The present study attempts to assess the impacts caused by relocation of villagers who were displaced involuntarily for the construction of Kotmale Dam – resettlers falling under category (2) above – in newly created agricultural farmland, by appraising their level of satisfaction more than 25 years after their relocation. It is assumed that such a study would complement existing relocation studies by providing a reasonable, longer-term assessment of the socio-economic impacts of population displacement due to infrastructural development. It is also expected to supplement the earlier findings by Takesada, Manatunge, and Herath (2008), who stated that resettlers' satisfaction is based on perceptions that reflect different strategies for coping with involuntary resettlement, and those of Manatunge et al. (2009), who described how providing strategic alternative economic opportunities will lead to satisfaction if resettlers can successfully rebuild their livelihoods.

There are numerous studies that report whether a particular resettlement scheme has been successful, which sometimes gives a wrong impression about the performance of the scheme if only basic socio-economic data such as changes in income levels or ownership of assets are analyzed. Satisfaction or dissatisfaction of resettlers cannot always be related to materialistic benefits; the reasons behind those perceptions have to be carefully analyzed to come to conclusions about the success of involuntary resettlement schemes. Therefore, the present study specifically analyzes the responses provided by resettlers affected by the Kotmale Dam as to their current status compared with that before resettlement and their overall satisfaction with the scheme. This paper attempts to explore the nature of compensation packages provided to resettlers and the material means through which they attempted to restore quality of life and provide satisfaction.

Mahaweli Development Project and Kotmale Dam

The Kotmale project is one of five major headworks projects that were undertaken under the Mahaweli Development Project. Financially assisted by the Government of Sweden, Kotmale Dam is the furthest upstream of the projects and was developed to regulate river flows in addition to harnessing the hydropower potential of a major right-bank tributary of the Mahaweli River, the Kotmale Oya. The Kotmale Oya flows through the rural mountain regions of Sri Lanka, passing ancient villages steeped in history and tea plantations of a more recent era.

The Mahaweli Development Project covers approximately 210,000 ha of farmland, where nearly a million persons were settled during 1970s and 1980s, including resettlers from the four dams (Victoria, Randenigala, Rantembe and Kotmale) constructed under the project. The command area has been divided into 'Mahaweli systems' (sub-regions), where Systems A through G are contiguous regions located on the lower reaches of the Mahaweli (Figure 1). These areas are in the dry zone, which covers almost three-quarters of the country and is characterized by a long dry season, high annual rainfall variability, and warm climate. To improve the quality of life of resettlers of Mahaweli Development Project (both voluntary and involuntary), a well-planned physical, social, institutional and economic infrastructure was provided in these newly created resettlement schemes (Mahaweli systems).

Resettlement options for Kotmale Dam

From the late 1970s to the early 1980s, the Kotmale reservoir (one of the five projects under the Mahaweli Development Project) flooded nearly 4000 ha of fertile land in the Mahaweli upper catchment, which included about 600 ha of paddy fields and caused the resettlement of 3056 families due to inundation.

Two options were provided for the displaced.

Option 1: Agricultural land (2.5 acres of irrigated dry land and/or rice fields and 0.5 acres for the home plot) from the new Mahaweli Systems B, C, and H (Figure 1). The 1722 families who selected this option were allowed to choose which of the systems to move to.

Option 2: Tea plots near Kotmale Reservoir, a scheme selected by 1334 families, which were resettled in 17 settlements around the reservoir. The size of the compensation tea plots was determined based on productivity: 0.51 ha (1.25 acres) of low-producing seedling tea or 0.3 ha (0.75 acres) of vegetatively propagated higher-yielding tea allotments.

Both communities received similar compensation packages in terms of economic returns. Though falling short of best-practice guidelines, the provisions did attempt to alleviate some of the hardships of relocation to ensure that people benefited from the development.

Resettlers were initially allowed to choose between the two options but could not revisit their choice later. The authorities delayed handing out the legal titles for the land, which was intended to prevent speculators from prompting resettlers to make hasty sales and depart. Substantial investments were made in technical assistance and training in agriculture-related activities, with the objective of protecting such new land titles at least until the resettlers had established new production systems and could make informed judgments about likely earnings.

The study's approach

This study's findings are based on several field visits to the resettlement sites, in which interviews were held with households using a structured questionnaire. Key informants and officials were interviewed separately.

Two household surveys were conducted: one in 2005 (Takesada et al., 2008), and another in 2011. The surveys covered 266 and 171 households, respectively (Table 1).

For any kind of post-project review, the assessment of social impacts can follow the 'impoverishment model' as suggested by Cernea (2000). In this model there are eight forms of impoverishment: landlessness; joblessness; homelessness; marginalization;

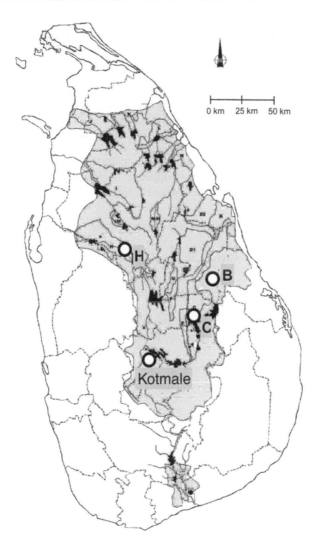

Figure 1. Location of resettlement sites in Sri Lanka. Shaded areas represent land covered by the Mahaweli Development Project. *Source:* Planning & Monitoring Unit, Mahaweli Authority of Sri Lanka.

increased morbidity and mortality; food insecurity; loss of access to common property; and social disarticulation. These parameters can be used to identify the strengths of the resettlement scheme, to review and assess the negative impacts, and to propose measures for prevention or mitigation.

The questionnaire for Survey 1 comprised the following elements:

1. Biographical and personal attributes of respondents
2. Economic activities to measure levels of satisfaction; opportunities created or lost after resettlement
3. Respondents' satisfaction with respect to physical or material well-being
4. Health status of resettlers; improvement or deterioration of household sanitation, water supply, etc.; education and related issues

Table 1. Number of households in the surveys.

Location of resettlement	Survey 1	Survey 2
Vicinity of reservoir	70	34
System B	64	42
System C	54	49
System H	78	46
Total	266	171

5. Questions to measure self-assessment and perceptions regarding the resettlement scheme, including self-esteem; evaluation of new opportunities and general satisfaction; review of the compensation scheme to identify resettlers' views

The questionnaire for Survey 2, aimed at collecting further details, included:

1. Economic conditions (trends in the household economy); perceptions of the community
2. Assessment of satisfaction regarding irrigation and water availability; roles of farmer organizations
3. Income-generating activities (on-farm, off-farm and non-farm)
4. Community participation and institutional support in rebuilding social infrastructure
5. Improvement or impoverishment of quality of life, social problems, safety of people with respect to human–elephant conflicts, etc.

The two authors, together with four local field assistants, carried out the interviews, during which the first author interpreted and translated the questions and responses.

Household surveys that seek clues about satisfaction are difficult to administer or interpret because respondents often lack a context for or explanation of the questions; therefore, they often skip questions or provide vague responses. In the surveys presented here, attitudes were appraised indirectly, by asking respondents about their reactions to hypothetical situations – for instance, by asking them what they would opt for if they were given a choice of the place of resettlement now.

Data on the cropping patterns and incomes of prospective resettlers in the pre-settlement villages are scanty. General profiles for traditional rural families are available, but not for Kotmale. We asked the interviewed householders to compare present earnings from different sources with income levels before they resettled. Few quantitative references to pre-settlement net incomes refer to wage earnings as labourers or sharecroppers, or cash sales of crops. The omission of subsistence products consumed at home is noteworthy. Such goods, including firewood, grazing fields, etc., were the basis for the stable household economy many resettlers have lost. Reflections on pre-settlement lifestyles among those who are involuntarily moved always tend to exaggerate the agreeable conditions of yesteryear (World Bank, 1998). Even so, comparisons of pre- and post-dam cash incomes do not reflect the benefits of the earlier lifestyle that now matter most to the resettlers (World Bank, 1998).

Repeated meetings with a small number of households, supplemented by interviews with key informants and community groups, is an effective low-cost technique for tracking the performance of rural development projects (World Bank, 2004) such as the highlighted case study presented in this paper.

Observations

General levels of satisfaction

People who were resettled in the vicinity of the reservoir, i.e. the Kotmale area, were more satisfied than those who had moved to Mahaweli Systems B, C and H. (Results of the questionnaire survey are shown in Table 2.)

More than 95% of those who were settled in Kotmale were satisfied with the resettlement option, whereas rates among those who had moved to Mahaweli systems was around 70%. A small proportion of people, especially in Systems C and H, replied that they would be satisfied provided that certain conditions were fulfilled, such as the availability of more irrigation water and more options for the second generation.

To check whether the responses of the resettlers were consistent and reliable, we asked a hypothetical question (at the end of the questionnaire): what their choice would be if they were given the chance to select the place of resettlement now. The responses obtained are summarized in Table 3.

The number of households who were satisfied with the choice of resettlement (as indicated in Table 2) closely matches the responses given for the choice of resettlement if made now (Table 3). This is an indication of the dependability of the responses given by the households.

Reasons for selecting the resettlement scheme

The majority of the households – about 80% of those who moved to System C, and more than half of those who moved to the other two systems – wanted to move to Mahaweli

Table 2. Household satisfaction with respect to choice of resettlement.

Are you satisfied with the choice you made in selecting the resettlement option?	Kotmale		System B		System C		System H	
	Households	%	Households	%	Households	%	Households	%
Yes	67	95.7	47	73.4	38	70.4	56	71.8
No	3	4.3	17	26.6	16	29.6	22	28.2
TOTAL	70	100	64	100	54	100	78	100

Table 3. Choice of the resettlers, given another chance. (Numbers in parentheses represent the number of households who expressed their satisfaction with selection of the particular resettlement option given in Table 2.)

If you were given a chance, would you make the same choice now?	Number of households			
	Kotmale	System B	System C	System H
Yes	69 (67)	45 (47)	34 (38)	56 (56)
No	1 (3)	19 (17)	20 (16)	22 (22)
TOTAL	70	64	54	78

systems because they wanted to continue paddy cultivation (Table 4a). The second noteworthy reason was the influence of relatives who had decided to move to these areas, which indicates that some households made collective decisions to move to the same area.

The numbers in parentheses in Table 4a represent the number of households who expressed their satisfaction in selection of a particular resettlement option, results which were obtained from the questionnaire survey. It is clear that the majority of those who chose paddy cultivation as the reason for choosing the resettlement option are generally satisfied. For example, 35 of 37, 32 of 43, and 39 of 41 households who decided to move to Systems B, C, and H, respectively, in order to continue paddy cultivation are satisfied with their choice of resettlement. In contrast, the number of satisfied households was smaller for those who moved for other reasons.

Those who wished to settle near the reservoir decided to do so because they did not want to move to Mahaweli systems due to the harsh climate of the dry zone of Sri Lanka (Table 4b). Also, they wanted to continue to live in Kotmale for reasons such as a wish to live in their ancestral villages, and less willingness to leave traditional villages whose locations are conveniently located close to established urban centres.

Traditionally, farmers from the Kotmale area have relied on well-adapted wet-rice cultivation along river valleys. Those resettlers who decided to stay in Kotmale faced a problem in being relocated high up in the hills. Given the total change of terrain and the insufficient water, they had to abandon wet-rice cultivation and start as small-scale tea cultivators. In fact their new land plots were subdivisions of unproductive tea estates. Agricultural extension services to care for this new group of tea growers and especially their need for training seems to have been almost non-existent. Resettlers in the Mahaweli systems, although they were able to continue growing rice, had to cope with a new

Table 4a. Reasons for selecting the resettlement option for those who moved to Mahaweli systems. (Numbers in parentheses represent the number of households who expressed their satisfaction with selection of the particular resettlement option given in Table 2.)

Reason for choosing the resettlement option	System B		System C		System H	
	Households	%	Households	%	Households	%
Preferred paddy cultivation	37 (35)	57.8	43 (32)	79.6	41 (39)	52.5
Received more land	2 (2)	3.1	1 (1)	1.9	3 (1)	3.8
Relatives moved	20 (9)	31.2	7 (3)	12.9	14 (6)	17.9
Did not want to grow tea	0	0	3 (2)	5.6	9 (6)	11.5
Had no option	4 (0)	6.3	0	0	5 (1)	6.4
Other reasons	1 (1)	1.6	0	0	6 (3)	7.6
Total	64		54		78	

Table 4b. Reasons for selecting the resettlement option for the vicinity of the reservoir.

Reason for choosing the resettlement option	Households	%
Had knowledge about tea cultivation	9	12.8
No family labour	6	8.6
Climate	45	64.3
Wanted to be at Kotmale	10	14.3
Total	70	

situation implying increased market integration and commercialization of the whole agricultural sector. This in effect meant that they found themselves changed overnight from small-scale, mixed-cropping, subsistence-oriented peasants to farmers producing a cash crop based on capital-intensive technology (Søftestad, 1991).

Reasons for satisfaction or dissatisfaction

Level and stability of income

The levels of income since resettlement have increased for nearly 60% of the resettlers in all four resettlement areas (Kotmale and Systems B, C and H) (Table 5). However, the stability of income has decreased after resettlement for more than 50% for those who moved to Mahaweli systems, whereas it has increased for 63% for those who resettled near Kotmale.

The satisfaction of the choice of resettlement has a direct relation with the increase of income levels. Almost all the households whose income increased after resettlement are satisfied with their choice of resettlement, which is seen by the numbers given in brackets (Table 5). This observation also is true in connection with stability of income. When the stability of income is increased, the resettlers are highly likely to be satisfied with their choice of resettlement, which is one of the reasons why those who resettled near Kotmale are more satisfied than those who moved to Mahaweli systems. During informal interviews, the key informants reiterated this: more stable income is as important as increased levels of income. For those who resettled near Kotmale, both the levels and stability of income increased, leading to relatively higher levels of satisfaction.

Land ownership

The extent of land ownership increased for more than 60% of all resettlers (Table 6). This again corresponds well with their level of satisfaction. The households who had a smaller land extent after resettlement, especially among the households who moved to System H, were not satisfied with their choice, mainly because of the decreased land ownership.

Table 5.   Change in levels of income and stability of income. (Numbers in parentheses represent the number of households who expressed their satisfaction with selection of the particular resettlement option given in Table 2.)

	Kotmale		System B		System C		System H	
	Households	%	Households	%	Households	%	Households	%
Level of income								
Increased	41 (41)	59	38 (35)	59	31 (29)	57	47 (44)	60
Decreased	7 (4)	10	10 (2)	16	9 (4)	24	15 (2)	19
No difference	22 (22)	31	16 (10)	25	14 (5)	19	16 (10)	21
Total	70 (67)	100	64 (47)	100	54 (38)	100	78 (56)	100
Stability of income								
Increased	44 (44)	63	31 (30)	33	19 (19)	35	20 (20)	26
Decreased	19 (16)	27	25 (10)	53	29 (15)	54	39 (22)	50
No difference	7 (7)	10	8 (7)	14	6 (4)	11	19 (14)	24
Total	70 (67)	100	64 (47)	100	54 (38)	100	78 (56)	100

Table 6. Change of land extent among the resettlers. (Numbers in parentheses represent the number of households who expressed their satisfaction with selection of the particular resettlement option given in Table 2.)

	Kotmale		System B		System C		System H	
	Households	%	Households	%	Households	%	Households	%
Increased	45 (42)	64	42 (37)	66	37 (32)	69	48 (43)	62
Decreased	20 (20)	29	15 (7)	23	12 (4)	22	25 (9)	32
No difference	5 (5)	7	7 (3)	11	5 (2)	9	5 (4)	6
Total	70 (67)	100	64 (47)	100	54 (38)	100	78 (56)	100

Table 7. Satisfaction with regard to irrigation water availability.

How satisfied are you with availability of water in your farmland?	System B ($N = 42$)		System C ($N = 49$)		System H ($N = 46$)	
	Soon after resettlement	Now	Soon after resettlement	Now	Soon after resettlement	Now
Satisfied	54%	78%	45%	81%	61%	85%
Somewhat satisfied	15%	13%	12%	5%	6%	8%
Not satisfied	31%	9%	43%	14%	33%	7%

Availability of irrigation water

The majority of resettlers were satisfied with the irrigation water availability, which included attributes such as quantity available, timely supply of water and overall management of the system by farmer organizations (Table 7).

The majority of resettlers are satisfied with irrigation water availability at present, though the majority had been unsatisfied soon after resettlement. Subsequent developments in irrigation water management, especially the input by farmer organizations, have been crucial in shaping the levels of satisfaction of resettlers. These increased availability of irrigation water will lead to successful paddy cultivation, together with higher income patterns, which will ultimately lead to overall satisfaction of the resettlers. In contrast, one area with which most respondents were dissatisfied was assistance with paddy marketing and the selling of their produce at reasonable prices, though every year the government offers a guaranteed minimum price to buy paddy.

Discussion

Of particular interest in this case study is why more of the the households who resettled in the vicinity of the reservoir ($>95\%$) are satisfied with their resettlement arrangement, compared to those who moved to Mahaweli systems (around 70%), despite receiving similar compensation packages in terms of income generation. Moreover, land ownership increased for those who moved to Mahaweli systems, with new infrastructure facilities to rebuild the social framework.

Increased income patterns and land ownership were common factors for satisfaction; however, income instability thwarted this sense of improvement. Households who resettled in the vicinity of Kotmale were more satisfied, and therefore it is apparent that the satisfaction levels among resettlers largely depend on circumstances other than what have been stated above. Two comparisons can be made:

1. Compare the proportion of satisfied households between households who resettled in the vicinity of the reservoir (Option 1) and those who moved to Mahaweli systems (Option 2).
2. Compare the proportion of satisfied households between those who selected different locations under Option 2, whether System B, C or H.

It was observed that education opportunities for the second generation are greater for Kotmale resettlers compared to those who moved to Mahaweli systems thanks to better facilities for secondary and higher education. The percentage of persons completing education up to secondary levels has increased since resettlement for all the locations (Takesada et al., 2008). Nearly 9% of those resettled in the vicinity of the reservoir and 3% of those who settled in System H had the opportunity to obtain higher education. In contrast, no resettlers in Systems B or C have obtained higher education. This extra opportunity for their children to obtain higher education has led to satisfaction with their locations of resettlement (Takesada et al., 2008). Kotmale and System H are located with easy access to major townships where there are well-equipped schools. It appears that, in moving to Systems B and C, the resettlers had to sacrifice – probably unknowingly – education opportunities for their children. The other reason that can be highlighted is the priority given to land ownership over other material benefits, which can be related to levels of poverty, landlessness and unavailability of information as to how they should plan their future lives. Without production assets to be passed on to their dependents, the parents were left with little choice in supporting the second generation, as described in detail by Takesada et al. (2008) and Manatunge et al. (2009).

In addition to abovementioned observations, the following are other reasons identified from interviews as to why the resettlers who moved to Mahaweli systems were not satisfied with their choice of resettlement location:

1. Harsh climatic conditions in the dry zone, compared to the mild conditions prevalent in Kotmale
2. Increased incidence of deceases, such as malaria, kidney disease, and high blood pressure (ascertained only from household responses), which had led to impoverishment, both economically and socially
3. Human–elephant conflict, which led to much destruction, severe stress among farmers and wasted time and money in protecting their farmland and produce

Attitudes differ, and there is no consensus between resettlers as to what leads to satisfaction: better productive farmland with irrigation water and fertilizer subsidy; more accessible schools and health facilities; off-farm income sources; etc. Judging by the actions of the families and their answers during interviews and other discussions, it is clear that some resettlers were satisfied despite insecure and unstable levels of outcome. Conversely, a small number of respondents said that their livelihood conditions and prospects for better economic outlook had deteriorated since resettlement. Yet, with some exceptions, they agreed that their homes were better built and farms better irrigated since resettlement.

The observations of this study show that negative consequences of resettlement are often offset by favourable factors such as access to irrigation infrastructure and other institutional support. This follows a similar observation to that reported for aquaculture development, for which proper management of aquatic resources is the key to maximization of benefits (Manatunge et al., 2009). Some farmers who had lost land to resettlement had valid grounds to reject the favorable comparisons between their present and past conditions that show satisfactory and stable incomes after resettlement. They claimed that although the whole Mahaweli program remained stable, they were not happy, because their income levels and stability of income had both decreased. In addition, harsh climatic conditions to which the resettlers are averse had led to more frequent health-related problems, thus affecting their physical capability, which determines whether they are fit and capable enough to engage in farming, share cropping or working as labourers. Most households who did not have family labour had to depend on hired labour, which had led to deterioration of their profits, thereby rendering their farming activities unprofitable (Takesada et al., 2008; Manatunge et al., 2009).

The primary rationale for the Mahaweli Development Programme, rice import substitution, ceased to be tenable shortly after the resettlement programme was commissioned back in the early 1980s because of a sharp drop in world rice prices (World Bank, 2004). This unquestionably led to decreased economic viability for rice cultivation, which has been the only source of income for most of the resettlers who moved to Mahaweli systems. Diversifying out of paddy farming was not realized for so long because the policy environment and the mind-set of Mahaweli officials were not conducive at that time, a circumstance which remains largely true even today. The resettlers who moved to Mahaweli systems were accustomed to deriving income from off-farm sources (e.g., rental of equipment, hiring of labour) and non-farm sources (wage work, cottage industries, employment outside or abroad) before they were resettled. However, with resettlement they lost such additional income sources, until re-establishing such activities a decade or so later.

Resettlers from Kotmale were not accustomed to large-scale paddy cultivation. Most of them did not have such farm skills as irrigation water management and crop selection depending on water availability. Also, Mahaweli staff members were not trained to foster transfer of such skills. There was a strong tradition from the time of resettlement to focus on overall paddy production targets rather than market-driven demand for produce. These circumstances always led to gluts and lower market prices, which resulted in less stable income patterns and therefore dissatisfaction among resettlers.

Although the majority of resettlers decided to move to Mahaweli Systems due to their intention of paddy cultivation, the resettlers were not provided with full title to their land, which is one of the reasons they could not prosper. The title was not given because of a paternalistic concern that they would speculate with the land rather than cultivate it. However, this was counterproductive; farmers did not have any other property for obtaining loans to offer as collateral, which was a reason most of the resettlers were unsatisfied.

Irrigation water needs to be priced to reflect the sectoral need for financing to meet recurrent expenditure and capital recovery. Irrigation operation and maintenance costs need to be recovered from users if such large-scale projects are not to be an excessive burden on public finances. Mahaweli systems would have catered better to the needs of the resettlers had such a cost recovery system been in place.

Resettlers who settled in Kotmale after receiving tea plots did not face this scenario of declining prices for their produce. Their main difficulty was the unproductive tea plots they received and lack of capital for investing in replanting tea or purchasing regular stocks of fertilizer to enhance the productivity of their limited land holdings (Manatunge et al., 2009). Manure and inorganic fertilizers are relatively expensive, and present yields of tea plots and incomes are too low to justify large expenditures on these

Table 8. Levels of education before and after resettlement.

	No education	Primary education only	Up to secondary education	Up to higher education
Vicinity of the reservoir				
Before resettlement	33%	63%	4%	0%
After resettlement	4%	69%	15%	9%
System H				
Before resettlement	36%	61%	2%	<1%
After resettlement	2%	80%	14%	3%
Systems B & C				
Before resettlement	34%	58%	8%	0%
After resettlement	0%	77%	23%	0%

Source: Takesada et al. (2008).

purchased inputs. For many farmers, it was impossible to escape from this low-productivity trap.

Mahaweli irrigation resettlement schemes were well connected to other parts of the country with a network of well-planned roads. Access to these schemes has been continually improved and has recently been upgraded with the establishment of new city centres, providing further opportunities for enhanced the marketability of surplus crops.

A majority of the Kotmale resettlers were able to overcome the difficulties they faced at the time of resettlement, and managed to restore their livelihoods. Such success stories are available elsewhere in Sri Lanka: the most commonly cited examples are the Pimburettewa Scheme (1971) and the Victoria Dam Project (1984). But it is difficult to assess how resettlers were affected in most of the cases because there are no or little data available on the situation prior to evacuation.

Conclusions

More farmland and increased income were some of the factors that contributed to satisfaction among the farmers who moved to Mahaweli systems. It appears that those resettled in the vicinity of the reservoir were more satisfied than those in Mahaweli systems, mainly thanks to more opportunities for the second generation (e.g. better opportunities for education) and higher income stability. Thus, it is important to give priority to the needs of second and future generations for sustainability of any resettlement scheme.

A majority of the households who moved to Mahaweli systems wanted to continue paddy cultivation as their main source of income. Land extent and water availability was satisfactory, but the income of resettlers was not stable due to market price fluctuations that made them develop negative perceptions about resettlement. Therefore, this shows that satisfaction in the choice of resettlement has a direct relation to stability of income.

Another factor in the resettlers' satisfaction was better irrigation infrastructure and water availability, which included attributes such as quantity available, timely supply of water and overall management of the system by farmer organizations. This implies that land-based compensation should always be supplemented by appropriate irrigation infrastructure, which will assure success of the resettlement scheme.

Other factors such as harsh climatic conditions, increased incidence of diseases and conflicts with wildlife also led to difficulties after resettlement, which caused much discomfort among resettlers. Success of any resettlement scheme depends on such secondary factors too; they should be eliminated to the extent possible as they will indirectly lead to lower quality of life, including economic hardship.

Diversifying out of paddy cultivation during the periods when rice prices continued to fall would have made the resettlers gain more from resettlement compensation. Therefore, diversification and introduction of alternative sources of income, sustainability of production capacity and economic viability in the long term are essential considerations in resettlement planning which may guarantee the establishment of livelihoods and stability of income.

Awarding land titles on a timely basis would have provided opportunities for resettlers to secure loans. However, awarding such legal title would also have led to land speculation, which can be counterproductive and leave the resettlers landless. In any case, livelihood-rebuilding efforts should be complemented with opportunities for securing financial assistance and access to credit, which is crucial in the success of any resettlement scheme.

Acknowledgements

The research carried out for this study was funded by the Mitsui & Co., Ltd, Environment Fund. This study was also partly supported by KAKENHI (24310189) and the Core Research for Evolutional Science and Technology (CREST) programme of the Japan Science and Technology Corporation.

References

Cernea, M. M. (1995). Understanding and preventing impoverishment from displacement: Reflections on the state of knowledge. *Journal of Refugee Studies, 8*(3), 245–264.

Cernea, M. M. (2000). Risks, safeguards, and reconstruction: A model for population displacement and resettlement. *Economic and Political Weekly, 41*, 3659–3678.

Eriksen, J. H. (1999). Comparing the economic planning for voluntary and involuntary resettlement. In M. M. Cernea (Ed.), *The economics of involuntary resettlement* (pp. 83–146). Washington, DC: World Bank.

De Wet, C. J. (2006). *Development-induced displacement: Problems, policies and people*. New York: Berghahn Books.

Manatunge, J., Nakayama, M., & Contreras-Moreno, N. (2001). Securing ownership in aquaculture development by alternative technology: A case study of the Saguling Reservoir, West Java. *International Journal of Water Resources Development, 17*, 611–631.

Manatunge, J., Takesada, N., Miyata, S., & Herath, L. I. (2009). Livelihood rebuilding of dam-affected communities: Case studies from Sri Lanka and Indonesia. *International Journal of Water Resources Development, 25*, 479–489.

McCully, P. (2001). *Silenced rivers: The ecology and politics of large dams*. London: Zed Books.

Mejia, M. C. (2000). Economic recovery after involuntary resettlement: The case of brick makers displaced by the Yaciretá Hydroelectric Project. In M. M. Cernea & C. McDowell (Eds.), *Risk and reconstruction* (pp. 144–164). Washington, DC: World Bank.

Miyata, S., & Manatunge, J. (2004). Lessons for sound policies in water resource management: Evidence from households' decisions towards aquaculture in Indonesia. *International Journal of Water Resources Development, 20*, 523–536.

Partridge, W. L. (1989). Involuntary resettlement in development projects. *Journal of Refugee Studies, 2*, 373–384.

Pearce, D. W. (1999). Methodological issues in the economic analysis for involuntary resettlement operations. In M. M. Cernea (Ed.), *The economics of involuntary resettlement* (pp. 50–82). Washington, DC: World Bank.

Robinson, C. W. (2003). *Risks and rights: The causes, consequences, and challenges of development-induced displacement*. Occasional paper. Washington, DC: Brookings Institution/SDAIS Project on Internal Displacement.

Scudder, T. (n.d.). *A comparative survey of dam-induced resettlement in 50 cases*. Retrieved from http://www.hss.caltech.edu/~tzs/50%20Dam%20Survey.pdf

Søftestad, T. (1991). Anthropology, development, and human rights: The case of involuntary resettlement. In E. Berg (Ed.), *Ethnologie im Wiederstreit. Kontroversen über Macht, Geschäft, Geschlecht in fremden Kulturen* (pp. 365–387). München: Trickster.

Takesada, N., Manatunge, J., & Herath, I. L. (2008). Resettler choices and long-term consequences of involuntary resettlement caused by construction of Kotmale Dam in Sri Lanka. *Lakes & Reservoirs: Research & Management, 13*(3), 245–254.

World Bank (1998). *Recent experiences with involuntary resettlement: Brazil-Itaparika*. Report No. 17544, Operations Evaluation Department. Washington, DC: World Bank.

World Bank (2001). *World Bank operational mannual: Operational policies. Involuntary resettlement*. Washington. DC: World Bank.

World Bank (2004). *Third Mahaweli Ganga Performance Report, Sri Lanka. Project performance evaluation report, Sector and Thematic Evaluation Group*. Washington, DC: World Bank.

World Commission on Dams (2000). *Dams and development: A new framework for decision-making*. London: Earthscan.

Atatürk Dam resettlement process: increased disparity resulting from insufficient financial compensation

Erhan Akça[a], Ryo Fujikura[b] and Çiğdem Sabbağ[a]

[a]Adıyaman University, Technical Programs, Turkey; [b]Faculty of Humanity and Environment, Hosei University, Tokyo, Japan

A survey of 99 resettled families displaced by construction of the Atatürk Dam in Turkey revealed that only a few of the families agreed to the resettlement plan and most of them resettled reluctantly. The compensation for this displacement was primarily monetary; however, the actual amount of the compensation did not reflect the market price of the land and most of the families presently own less land than they did prior to the resettlement. This resettlement adversely affected those who owned small parcels of land in particular, as many have stopped farming and are presently working as labourers or crop sharers. Many who owned large parcels of land were able to continue farming. The insufficient compensation offered by this project widened the disparity between these two groups.

Introduction

Dam construction in Turkey dates back to the 1930s following the foundation of the republic. At this time some small-scale dams were constructed that did not require a large population resettlement. Construction of the Keban Dam on the Euphrates, which began in the late 1960s, changed this trend. Turkey was now confronted with many resettlement problems (Batukan, 1969). Since the General Directorate of State Hydraulic Works (DSI) acquired 520,000 hectares of private land and 200,000 hectares of government-owned land in 1958 for development projects (Akyürek, 2005), a total of 1076 dam projects have been completed; a total of 356,327 people had been resettled as of 2005 (see Table 1) (Akyürek, 2005; Özbaycı & İçten, 2005). The Turkish government increased the number of individuals resettled since the 1980s with the Southeastern Anatolia Project (hereafter, GAP Project, the Turkish acronym); however, the socio-cultural and economic approaches of the 1990s have not been satisfactory for those resettled. Neither the compensation money nor the construction of resettlement sites has sufficiently provided all of their needs. In the case of the Atatürk Dam, the largest dam in the country, the government's ability to provide compensation did not meet the demands of those resettled. This study presents the findings of a survey concerning the socio-economic situation, 20 years after resettlement, of the families resettled as a result of the construction of the Atatürk Dam. This study will also discuss the lessons learned from this long-term post-project evaluation.

Table 1. The major dam resettlement issues in Turkey.

Dam	Year	River	Affected settlement sites	Affected people
Keban	1974	Euphrates	174	30,000
Karakaya	1987	Euphrates	105	45,000
Atatürk	1992	Euphrates	143	55,300
Çat	1996	Abdülharap	7	4,000
Dicle	1997	Tigris	20	2,878
Batman	1999	Tigris	17	10,858
Tahtalı	1999	Tahtali	6	7,331
Birecik	2000	Euphrates	4	1,260
Other dams			Approx. 600	200,000
Total			1,076	356,327

Sources: Özbaycı and İçten (2005), Akyürek (2005).

Resettlement caused by construction of the Atatürk Dam

The Atatürk Hydroelectric Dam is located in the south-eastern region of Turkey and was built to meet the country's increasing energy and water demands within the GAP Project (Figure 1). This dam supplies water to the GAP Project plains of Harran, Suruç and Samsat. The GAP project covers nine provinces with a total population of 10 million people. The GAP Project was expected to generate 27 billion kWh of hydropower in 2012 and to irrigate 1.8 million hectares of land, almost one-fifth of the irrigable land in Turkey, by 2010. The irrigation network will include 22 dams in all, with 19 hydropower plants and a total installed capacity of 7500 MW. The Atatürk Hydroelectric Dam has acted as an "engine" for economic development in the region, which was historically underdeveloped (Altinbilek & Tortajada, 2012).

The resettlement as a result of the construction of the Atatürk Dam began in 1989 and ran until 1991. The dam was put into service in 1992 with 2400 MW of power and 8900 GWh annual power production.

The land and homes of approximately 55,300 people were fully or partially inundated as a result of this construction; 1 town and 11 villages were fully inundated and 3 towns and 79 villages partly inundated. Out of 55,300 people, 19,264 from 3251 households

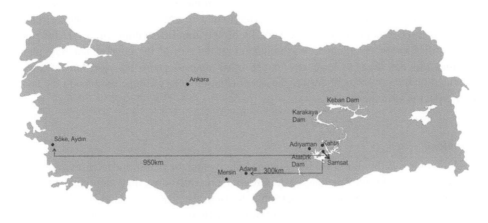

Figure 1.   Location of Atatürk Dam and other major dams on the Euphrates River, with resettlement locations.

resettled to New Samsat, Hatay, Söke and Aydın, where new resettlement areas were developed as a part of the governmental project. Others moved to Adıyaman and Kahta, which are close to the inundated settlement sites, as shown in Figure 1 (Akyürek, 2005). Others took their compensation and purposefully moved to Adana and Mersin, some hundreds of kilometres away from their homes (Kadirbeyoğlu, 2009). These cities are relatively developed, with a high potential for employment, and have been accommodating immigrants (particularly agricultural workers) from south-eastern Anatolia since the nineteenth century (Toksöz, 2010). Those families that resettled in the cities are experiencing a culture relatively similar to that in their previous homes.

Methodology

The interviews were conducted using questionnaire sheets from November 2011 to February 2012 in the resettlement areas close to the Atatürk Dam, including New Samsat, Kahta and Adıyaman. Ninety-nine resettled families were interviewed. They were grouped into two classes for analysis based on their income levels. Group 1 consisted of 33 well-off farming families. Each of these farmers were landowners with an income of more than USD1000/month, which is the threshold level for a well-off standard of living at the time of interview. Group 2 consists of 66 families with an income level below USD1000/month. Many of these farmers are seasonal workers, crop sharers, or farmers with land smaller than the country's average size of 5 hectares at the time of resettlement (Ballı, 2010).

Results and discussion

Resettlement and negotiation

Of the 99 resettled families, 55 indicated that details of the resettlement were explained to them beforehand and 54 indicated that these details were explained either by a government official or the leader of the village. Of the 99 resettled families, only 18 indicated that they had negotiated (or had someone negotiate on their behalf) concerning the resettlement. Only 7 families readily agreed to the resettlement, while 75 accepted reluctantly and 17 did not accept the resettlement at all. None of the 99 resettled families was provided land as compensation, but all of them received cash from the government. All of the families within Group 1 had options other than moving to new resettlement areas because of their relatively high compensation; however, 47 of the families within Group 2 did not. The notification of resettlement was unilateral on the part of the government. The poor families did not seem to have any choice other than to go to the resettlement areas provided by the government despite the insufficient monetary compensation and the lack of a proper resettlement plan.

Farmland

Table 2 compares the size of the farmland and family income before and after resettlement. While all of the families belonging to Group 1 presently own their farmland, only 51 (83%) of the families belonging to Group 2 do. Both the size of the land and the income decreased after resettlement except for a few large field owners. Because the income figure is nominal and does not account for inflation throughout this period of time, the income decrease was actually worse than the nominal figure.

Table 2. Median land size and family income.

Group	Land size (ha)		Income (USD/month)	
	Before	Presently	Before	Presently
1	26.0	14.0	2600	1900
2	1.4	0.7	500	400

Although dissatisfied with the amount of compensation, large field owners received a relatively high rate of compensation money from the government, which gave them a chance to select their resettlement sites because they had sufficient funds to establish new farms in many parts of the country. This was not an option for many of the small field owners due to the low compensation they received for their farm size. Prior to the resettlement, families expected to receive USD13,000 and USD25,000 per hectare of cereal land and pistachio nut orchard, respectively. The actual market price for such land at the time of resettlement varied from USD10,000 to USD13,000 (Gule Parlak, 2007). However, the compensation actually paid to the families was USD8,000 and USD10,000 per hectare, respectively, regardless of its condition – i.e. depth, slope, stoniness, drainage, texture, etc. (Guler Parlak, 2007). This is shown in Table 3. The government compensated the farmers on the basis of the real estate tax statements, which gave much less than the market prices (Demir, 2009). In fact, the results of the survey indicated that the actual compensation paid was about 40% to 60% less than expected. Moreover, the price of land in the newly developed resettlement areas was greater than expected because of land speculation. As a result, the majority of the resettled individuals had to accept less land and less family income than before the resettlement. While Altinbilek and Tortajada (2012) concluded that resettlers had generally been fairly compensated for their loss, they reported a significant delay in resettlement implementation. During the time lag between compensation and resettlement, land price might have further increased and made it much more difficult for the resettlers to buy new land.

In terms of irrigation, the situation was slightly better. Of the families interviewed, 16 (48%) and 50 (76%) of Group 1 and Group 2 farmers, respectively, indicated that they presently have sufficient irrigation. The relatively low satisfaction of Group 1 farmers is perhaps due to their experience of sufficient water supply prior to the resettlement. Group 2 farmers with relatively small land were most often close to the bank of the Euphrates, whereas Group 1 farmers with large fields were often located far from the Euphrates, prior to the resettlement. Following the resettlement, much irrigation infrastructure was developed by the GAP to allow for better irrigation opportunities for large field owners. Kapur, Kapur, Akça, Eswaran, and Aydın (2009) suggested a double or triple increase in crop yield following the new irrigation infrastructure. Altinbilek and Tortajada (2012) reported the gross agricultural output value of the GAP region after five years of irrigation (in 2000) at around USD262 million, indicating USD2347 per hectare and USD2547 per

Table 3. Compensation amounts at the time of resettlement (US dollars).

Land type	Expected	Received
Cereal field	13,000/ha	8,000/ha
Pistachio nut orchard	25,000/ha	10,000/ha
House with garden	40,000 per 250 m^2	20,000 per 250 m^2

Sources: Parlak (2007) and the authors.

capita. All of this reflects a net increase in income for the region. However, resettled families located in non-irrigated areas of the GAP region were unable to fully enjoy the benefit of this project.

Occupation

While most of the families in Group 1 have continued to farm, more than half of the self-employed farm families in Group 2 have changed their occupation (Table 4). Eleven resettled families from Group 2 have become labourers, whereas no family was engaged in labour prior to resettlement. This suggests that small landowners could not obtain sufficient land to maintain their lifestyle and have therefore been forced to change their occupation. To further support this idea, consider that 10 of the families in Group 2 indicated that they miss their farmland most desperately after resettlement. As a result, families in Group 2 have indicated a general dissatisfaction with their job, an increase from 6 to 49 after resettlement. Twenty (60%) and 45 (68%) families from Groups 1 and 2, respectively, felt that obesity increased after resettlement. The amount of self-produced food has decreased because they lost farmland or have smaller land since resettlement. They expressed that they had to purchase more food from the market and eventually changed their eating habits, consuming less vegetables. They felt that increasing obesity could be attributed to their changing eating habits.

Resettled families hope that their children pursue future careers as teachers or government officials (Table 5). While 14 resettled families from Group 1 hope their sons will continue farming, only 6 resettled families from Group 2 share this idea. Resettled families from Group 2 are small landowners and some have already abandoned farming; therefore, it is unlikely that they would want their sons to continue farming.

Table 6 reports what the resettled families want most from their government in addition to monetary compensation. The questionnaire asked the families to provide three answers according to their importance: first, second, and third. While the families from Group 1 requested information concerning how to spend their compensation money, families from Group 2 requested support for employment, implying that they are dissatisfied with their current job opportunities.

Infrastructure

The quality of life in terms of infrastructure, education, health, transportation, public facilities and household goods are generally improved since resettlement. Turkey's GDP

Table 4.   Occupations of the resettled families.

	Group 1		Group 2	
	Before	Presently	Before	Presently
Self-employed farmer	33	31	44	20
Crop sharer	0	0	9	11
Public sector	0	0	1	2
Private-sector employer	0	0	1	1
Labourer	0	0	0	11
Trader	0	0	0	6
Unemployed	0	0	0	3
Other	0	0	0	12

Table 5.   Future occupations desired for children.

	Group 1		Group 2	
	Son	Daughter	Son	Daughter
Farmer	14	0	6	0
Company employee	0	0	2	0
Teacher	14	27	19	43
Public official	2	1	31	15
Military service	1	0	4	0
No work	0	5	0	4
Other	0	0	0	0

per capita averaged an annual growth rate of 2.3% between 1990 and 2010 (UNDATA, 2012) and it is likely that the quality of life of the resettled families improved along with this economic development. Mud and brick houses were common in the homelands of resettled families prior to resettlement; they now enjoy more cement houses. However, the average size of a house with a garden decreased from approximately $300\,m^2$ ($100\,m^2$ indoor area) to $200\,m^2$ ($80\,m^2$ indoor area) for families in Group 2. Facilities and opportunities for education have significantly improved, and all of the resettled families indicate that they are happy with this improvement. However, 34 of the families in Group 2 complained that opportunities for employment for their children have worsened. This is consistent with their hope for the government to create employment opportunities, as shown in Table 6. As a result, the majority (89 of 99) of the resettled families feel that the places where they presently live are good in terms of education, while many of them (80 of 99 families) feel that their economic conditions worsened after resettlement.

Local sentiment

Hattori and Fujikura (2009) suggested that the local sentiment of families resettled after construction of a Japanese dam had to be considered to mitigate their emotional

Table 6.   What resettled families want from the government, by priority.

	Group 1			Group 2		
	1st	2nd	3rd	1st	2nd	3rd
Establish a factory and employ resettled families	1	6	8	29	14	8
Provide agricultural training	6	5	5	6	18	7
Provide information concerning how to spend the compensation money	12	8	10	6	1	8
Establish a cooperative organization, which regularly takes care of the compensation money	6	11	4	2	7	13
Provide vocational training	7	1	6	3	7	15
Help to find new employment	0	0	0	14	15	3
Visit often and listen to our concerns	0	0	0	4	1	8
Educate children that many families were forced to resettle in order to construct a dam	1	2	0	2	0	0
Help locate a new house	0	0	0	1	3	0
Establish a community house for the resettled families	0	0	0	1	0	0
Other	0	0	0	0	0	0

dissatisfaction. The sentiment of the people living in the vicinity of the Atatürk Dam seems as important as it was in Japan. A common issue (in addition to the lower income) among resettled families is fewer opportunities to visit their relatives as a result of their living in different resettlement sites. It is common for those living in this area to frequently visit their parents and relatives. Among the resettled families in Group 2, 65 of the 66 families visited their relatives at least once each month prior to resettlement; after resettlement, this number decreased to 40. A similar result was observed among the resettled families in Group 1 (a decrease from 21 to 14). All of the 33 families in Group 1 indicated that losing the opportunity to visit their parents and relatives was very upsetting. Such local sentiment does not seem to be as important among the second and third generation; however, it is an important issue among the first generation.

Resettled families also complained about their loss of social status as a result of resettlement. Owning a farm and a house is a symbol of prestige in south-eastern Turkey. Families in Group 1 were mostly landlords of large farms and thus more respected in their homelands. However, the local community of the new settlement sites did not take this social status into account. Families in Group 2 were disappointed that they had lost their farms and were sometimes considered refugees in the new resettlement areas.

Conclusions

Negotiations between the government and the resettled families do not seem to have been conducted appropriately. Only 7 of the 99 resettled families agreed with the resettlement plan and the others either agreed reluctantly or did not agree at all. The compensation was monetary, but the amount of money did not reflect the market price of their submerged land. Moreover, the land price of the resettlement areas increased because of land speculation.

As a result, it became increasingly difficult for small landowners to sustain their lifestyle by farming on their reduced land. Therefore, many were forced to change their occupation. Such families want the government to provide job opportunities such as factory employment, and the majority want their children to pursue jobs other than farming in the future. These families seem to be pessimistic about the future of farming. On the other hand, large landowners were able to continue farming despite the fact that they now have less land than prior to resettlement. These families want the government to provide advice about how to spend their compensation money rather than providing other job opportunities. More of these families want their sons to do farming. Therefore, resettlement seems to have increased the disparity between these small and large landowners.

While the property and educational opportunities of the resettled families have significantly improved thanks to the nation's economic development, resettled families are still generally dissatisfied with their resettlement. An important factor in such resettlements is emotion, such as in the case of Japan (Hattori et al., 2009) and Indonesia (Agnes, Solle, Said, & Fujikura, 2009). The families missed the opportunity to visit with their relatives and missed the social status they enjoyed in their home communities.

The government should have provided compensation based on the market price of the land. Moreover, if the government develops resettlement areas it should implement measures to avoid price increase due to land speculation and/or provide the land exclusively for the resettled families, especially in the case of small landowners. In the case of the Japanese dam construction, the government developed two resettlement areas and provided exclusively resettled families with a cheaper price (Hattori et al., 2009).

If such measures are not implemented, land-for-land compensation should be considered at the very least. Few of the small landowners want their children to continue farming, and therefore vocational training and industrial development will be key for the future generation.

Acknowledgements

This study was funded by the Mitsui & Co., Ltd, Environment Fund and KAKENHI (24310189). The authors would also like to thank Ms Fatos Ercamal and Ms Nazli Tanir for their kind support during field surveys.

References

Akyürek, G. (2005). *Impact of Atatürk Dam on social and environmental aspects of the Southeastern Anatolia Project*. Unpublished Ph.D. thesis. Middle East Technical University, Ankara, Turkey.

Agnes, R. D., Solle, M. S., Said, A., & Fujikura, R. (2009). Effects of construction of the Bili-Bili Dam (Indonesia) on living conditions of former residents and their patterns of resettlement and return. *International Journal of Water Resources Development, 25*, 467–477.

Altinbilek, D., & Tortajada, C. (2012). The Atatürk Dam in the context of the Southeastern Anatolia (GAP) Project. In C. Tortajada, D. Altinbilek & A. K. Biswas (Eds.), *Impacts of large dams: A global assessment* (171–199). Berlin: Springer.

Ballı, B. (2010). The landownership in Turkey. *Turkish Agriculture Journal, 192*, 23–29, [in Turkish].

Batukan, I. (1969). Housing problems of Keban Dam. *Bulletin of Engineers, 11*, 16–19.

Demir, M. (2009). Factors affecting tax evasion. *Journal of Justice, Economy and Social Sciences* [in Turkish]. Retrieved from http://e-akademi.org/makaleler/mdemir-3.htm

Guler Parlak, Z. (2007). *Dance of life with water: Dams and sustainable development*. Ankara: Turhan [in Turkish].

Hattori, A., & Fujikura, R. (2009). Estimating the indirect costs of resettlement due to dam construction: A Japanese case study. *International Journal of Water Resources Development, 25*, 441–457.

Kadirbeyoğlu, Z. (2009). Case study Turkey. *Environmental Change and Forced Migration Scenarios (EACH-FOR) Newsletter*, No. 5, EU 6th Framework Project.

Kapur, S., Kapur, B., Akça, E., Eswaran, H., & Aydın, M. (2009). A research strategy to secure energy, water, and food via developing sustainable land and water management in Turkey. In H. G. Brauch, U. O. Spring, J. Grin, C. Mesjasz, P. Kameri-Mbote, C. N. Behera,... H. Krummenacher (Eds.), *Hexagon Series on Human and Environmental Security and Peace* (Vol. 4) (pp. 509–518). Berlin: Springer.

Özkalaycı, Z. E., & İçten, H. (2005). Resettlement planning and its applications at General Directorate of State Hydraulic Works., Turkish Chamber of Mapping and Cadaster Engineers *10th National Map Science and Technics Congress*, 28 March–1 April 2005, Ankara, Turkey.

Toksöz, M. (2010). *Nomads, migrants and cotton in the Eastern Mediterranean: The making of the Adana-Mersin Region 1850–1908*. Leiden: Brill.

UNDATA (2012). *GDP per capita average annual growth rate*. Retrieved from http://data.un.org/Data.aspx?d=SOWC&f=inID%3A93

The long-term implications of compensation schemes for community rehabilitation: the Kusaki and Sameura dam projects in Japan

Kyoko Matsumoto, Yu Mizuno and Erika Onagi

Department of International Studies, Graduate School of Frontier Science, University of Tokyo, Japan

Very few studies have been conducted to analyze the long-term consequences of large infrastructure development on community rehabilitation. This study reviews the Kusaki and Sameura dam projects in Japan, which were carried out in the 1970s. This research attempted to identify factors in the compensation schemes and resettlement negotiations of these projects that affected long-term community rehabilitation and individual resettlement. The lessons learned from this study will provide valuable knowledge for developing countries where large infrastructure development has been vigorously undertaken.

Introduction

A number of dam projects were undertaken in Japan during the reconstruction and economic development that took place after World War II. As demand for water, food and electricity increased, dam construction became a key factor driving economic growth. The government needed to address this urgent demand for increased resources.

After the war and until the early 1960s there was no general rule or standard for providing compensation for submerged properties (Takesada, 2009). With insufficient awareness and compensation measures, controversies between dam developers and local people concerning involuntarily resettlement were ubiquitous. Compensation measures such as the Guidelines on Standards for Compensation for Losses Associated with the Acquisition of Land for Public Purposes (*Koukyou Youchi no Shutoku ni tomonau Sonshitsu Hoshou Kijun Youkou*, enacted in 1962, hereafter referred to as the Guidelines) (Government of Japan, 1962a) were enacted when it came to public land acquisition. The main purpose of these guidelines focused on expediting construction projects rather than providing a smooth transition for affected people (Hanayama, 1969). Insufficient compensation measures for the public and individuals had no small effect on the falling population and community collapse.

This research was conducted near the Kusaki Dam at the village of Azuma, Gunma Prefecture, and near the Sameura Dam at the village of Okawa, the town of Tosa and Motoyama, Kouchi Prefecture, Japan. Both dams were constructed in the 1970s, which was a period of high economic growth for Japan. The full effects of the involuntary resettlement associated with these projects often spanned several decades; however, few studies have examined the long-term consequences of this resettlement. Therefore, this study focused on the long-term consequences of the compensation schemes offered through the dam projects. The research objectives were to capture valuable lessons from Japan's past experiences of involuntary resettlement and this article will focus on how the compensation schemes have influenced community rehabilitation and individual lives over a long period of time.

This study will attempt to provide some direction for the implementation of effective compensation schemes for infrastructural development in developing countries where large dam construction projects are currently moving forward. These developing countries face a very similar socio-economic environment to Japan in the 1970s, including rapid urbanization and a change in the lifestyles of people. The Japanese experiences may provide valuable lessons for these developing countries.

Compensation policies and systems of resettlement in Japan

The Guidelines stipulated principles of compensation for loss of land and property in financial terms only (Hattori & Fujikura, 2009). At the same time, the Memorandum on the Implementation of the Guidelines on Standards for Compensation for Losses Associated with the Acquisition of Land for Public Purposes (*Koukyou Youchi no Shutoku ni tomonau Sonshitsu Hoshou Kijun Youkou no Shikou ni tsuite*) (Government of Japan, 1962b) was passed at the cabinet level. This memorandum specified that only property rights were subject to compensation; but despite the fact that compensation for livelihood was denied, there was a provision for the restoration of livelihood, job placement or guidance where required.

To promote the restoration of dam-affected areas, the Guidelines on the Standards for Public Compensation Associated with the Implementation of Public Works Projects (*Koukyou Jigyou no Shikou ni tomonau Koukyou Hoshou Kijun Youkou*) were passed by the cabinet in 1967 (Government of Japan, 1967). The loss of public facilities, including schools, public offices, railroads and roads, was compensated under this standard. The principle behind public compensation was to rehabilitate the function of public facilities by monetary compensation. Although these guidelines and memorandum were based on the principle of monetary compensation, they still recognize other kinds of compensation if they are considered practical from a technical or economic standpoint (Hattori & Fujikura, 2009). After the enactment of these guidelines, the dam developer could officially provide compensation for the loss of building facilities, including schools and roads (Hattori & Fujikura, 2009).

Two laws were enacted focusing on livelihood rehabilitation: the Law Concerning Special Measures on Biwako Integrated Development Project (*Biwako Sougou Kaihatsu Tokubetsu Sochihou* – hereafter, Special Measures on Biwako) (Government of Japan, 1972) and the Law Concerning Special Measures in Water Resources Areas (*Suigen Chiiki Tokubetsu Taisaku Sochi Hou* – hereafter, LSM) (Government of Japan, 1973). The Special Measures on Biwako state that upper municipalities bear the expense of downstream municipalities. The LSM was enacted in 1973 to improve welfare and the stability of livelihoods in the affected area. The law states that relevant authorities are to provide resettled individuals assistance in obtaining land, buildings, job placement, guidance and training (Article 8). The law also stipulates that local governments shall endeavor to take the necessary measures in order to contribute to the revitalization of water resources areas (Art.14, LSM) (Government of Japan, 1973).

The Watershed Countermeasures Fund (*Suigen Chiiki Taisaku Kikin*) was established to promote livelihood rehabilitation and development in water resources area. The first

Fund was applied to the Tone and Arakwa rivers in 1976. The Fund serves as a financial mechanism to provide lower-interest loans (better-than-market loans) for dam-affected households as they transition to their resettlement sites, to provide livelihood consultation personnel, to build or improve roads for livelihood purposes, and to facilitate communication between upstream and downstream communities (Hattori & Fujikura, 2009). In addition, the Land Expropriation Law (*Tochi Shuuyou Hou*) (Government of Japan, 1951a) was revised in 2001 and now mandates livelihood restoration for resettled individuals in any area of public works. Since the enactment of the Guidelines, the level of compensation for livelihood rehabilitation has been gradually reshaped and has thus resolved the increasing conflicts between developers and resettled individuals.

The Kusaki Dam Project

Some 221 households were resettled as a result of the Kusaki Dam Project. Of these 221 households, 103 (about 46%) moved out of the village; 94 of these 103 households moved to neighbouring towns (Muzushigen Kaihatsu Koudan et al., 1974). The village of Azuma was the only community affected by the Kusaki Dam construction. The stone industry was the main industry present in the area before resettlements and continues to be to this day.

Resettlement negotiation

The Japanese government enacted the Multiple Purpose Land Development Law (*Kokudo Sougou Kaihatsu Hou*) (Government of Japan, 1950) in 1950, aiming to revitalize agriculture, flood control and industry after the war (Seta gun Azuma mura, 1998). In Gunma Prefecture, the Comprehensive Development Plan at Tone Specific Area (*Tone Tokutei Chiiki Sougou Kaihatsu Keikaku*, cabinet decision in 1951) (Government of Japan, 1951b) was designed to promote agriculture and electricity production as well as the forest industry. Under this plan, multi-purpose dam developments were planned in Gunma Prefecture from the late 1950s to the 1960s (Seta gun Azuma mura, 1998).

The Kusaki Dam was the first multi-purpose dam constructed on the Watarase River. After the pilot investigation for the dam construction was initiated by the Ministry of Construction in 1958, the affected communities increasingly raised protests against the dam construction. Among the affected communities, the Kusaki district was significantly inundated; residents of this area took the lead in establishing the Alliance for Construction Resistance with other affected districts in 1963 (Seta gun Azuma mura, 1998). Those involved in the Alliance were stone cutters, because the majority of resettled individuals were involved in the stone industry; their demands as a group were positive since this allowed them to continue working on quarrying.

At the same time, the Azuma village office launched a committee for the dam construction. This committee served as a focal point for the Water Resources Development Public Corporation (*Mizushigen Kaihatsu Koudan* – the former Japan Water Agency, hereafter WRDPC). The Village Assembly also established a special committee for dam construction, whose aim was to study the issues related to dam construction and to implement provisions (Seta gun Azuma mura, 1998).

The prefectural government and the Assembly initiated mediation talks, hoping for a breakthrough in the deadlock between the WRDPC and the villagers. Petitions from the Alliance were delivered to the prefectural government and the WRDPC via the village committees. According to the petition dated July 1966, the committee requested compensation for livelihood rehabilitation, including alternative land for homes, public compensation, promotion of local industries and tourist development (Seta gun Azuma mura, 1998). In response to these petitions, several joint field visits by the prefectural government were dispatched to Azuma. These structural negotiation units attempted to identify and consolidate villagers' requests. After many meetings between the village and the WRDPC the villagers changed their attitude from *completely opposed* to *opposed with conditions* in 1965. An investigation of land boundaries began in 1968. Prior to this investigation, the village asked the WRDPC to announce the provisional compensation rule and rehabilitation plan for the stone industry. The village also asked the WRDPC to set a minimum level of compensation for those in the village (Seta gun Azuma mura, 1998). The villagers also submitted a petition for livelihood rehabilitation, because construction related to the dam had already begun; villagers were concerned about the lack of provisions for livelihood rehabilitation despite the fact that the project had already started. The Council for the Countermeasure of Rehabilitation of Livelihood was established in 1968 to discuss resettlement measures. Throughout this consultation the WRDPC took the stance that negotiations for individual and public compensation would take place parallel to one another; however, the village office was strongly against such parallel negotiations because they believed that once the construction began their negotiation for public compensation would be less productive. These actions demonstrate how much the villagers were concerned about their compensation for livelihood rehabilitation.

After the Alliance and the WRDPC reached an agreement (which included an explanatory meeting for the compensation offered), the Kusaki Dam Project moved forward. The WRDPC began construction of the Kusaki Dam in 1971 after the conditions of the public and individual compensations were finally agreed upon.

Compensation schemes for the Kusaki Dam resettlement

Several compensation measures were carried out to facilitate the resettlement of those affected by the Kusaki Dam Project. The Alliance for Construction Resistance played a key role in the negotiation of these measures. The Alliance had significant input into the pre-construction aspects of the project, including the field visits to existing dams and provisions for rehabilitation. Throughout the process of discussion, the Alliance gradually identified their requests and set their target for compensation. According to an interview with a son of a core member of the Alliance, "the ultimate purpose of the Alliance was to gain compensation for a house, even for people who did not own a house or even lease a plot at the time" (interview by the authors, 2011). Also, the petition submitted to the WRDPC in 1968 inquired about the minimum level of compensation that would be given (Seta gun Azuma mura, 1998). The Alliance faced many difficulties throughout their negotiations but gradually achieved their objective of obtaining a minimum level of compensation for the weak (i.e., those who do not own enough property to sustain their livelihood).

An evaluation of material loss entails counting the items lost in exact detail, including, for example, trees; this method allows residents to claim the greatest possible loss. The resettled individuals interviewed by the authors were under the impression that they could receive cash compensation for their material loss in accordance with the cash compensation rule. Some individuals planted additional trees just prior to the evaluation so that they could receive more compensation than others around them. As a matter of fact, the appraisal of material loss was often inflated by an arbitrary amount even after the Guidelines were enacted (Maruyama, 1986). The individuals were focused on receiving their full compensation and rarely cared about the details of what they were being compensated for.

The national forest adjacent to Azuma was handed over to stone cutters for their stone resources. Financial support for driving licenses was also provided by the WRDPC to improve mobility and alternative transportation. In this way some resettled individuals could begin their own businesses as carriers in the transport industry. The bus service helped children and resettled individuals near dam sites to commute to schools and work places. Those interviewed indicated that these compensation measures were effective for securing jobs, conveniences, education, a future for children and mental stability for those who stayed near the dam sites.

The Sameura Dam Project

The Shikoku-Region Development Promotion Act (*Shikoku Chihou Kaihatsu Sokushin hou*) (Government of Japan, 1960) was put in place in 1960 for the purpose of developing natural resources in the Shikoku region of Japan. Among the projects associated with this initiative, the water resource development plans for the Yoshino River were regarded as core projects. The Sameura Dam was built in 1973 as a part of the water resource development plans for the Yoshino River. It was intended to supply large amounts of water to neighbouring prefectures, and its construction made a significant contribution to the economic development of the Shikoku region.

The areas affected by this project include the towns of Motoyama and Tosa and the village of Okawa. Among these, Tosa and Okawa were the main areas inundated by the dam reservoir. The resettled individuals from this area included 352 households (Mizushigen Kaihatsu Koudan, 1979). Although the submerged area of Okawa was smaller than that of Tosa, a central portion of Okawa, including government offices, schools and businesses, was submerged. Therefore, Okawa was most significantly affected by the construction; 141 out of 167 households moved to neighbouring towns or other cities, causing a sharp decrease in the population (Okawa mura, 1981). Furthermore, the mining industry (*Shirataki Kouzan,* the Shirataki Mine Cooperation) was the main industry in the village and included over 2000 workers. This industry was forced to shut down in 1971 due to trade liberalization that emerged after construction began. This further accelerated the exodus from the village and the population decreased from 3212 to 2206 at that time (Mizushigen Kaihatsu Koudan, 1979). Those who remained in the village attempted to redevelop the region and industry by engaging other local resources (e.g. forests and livestock); however, these efforts were complicated by the rapid depopulation and in some cases were abandoned.

Resettlement negotiation

The village of Okawa suffered the most significant destructive impacts from the dam project. In view of the anticipated impact, the villagers had strongly protested against the project. In 1962, the villagers from Okawa and Tosa jointly passed a resolution in opposition to the dam construction and established the Alliance for Construction Resistance (Okawa Mura Shi Tsuiroku Hensan Iinkai, 1984). In that same year, Okawa constructed a new public office within the projected dam site as a symbol of protest. The villagers also refused external investigation aimed at evaluating the worth of their properties and fields. The people of Okawa worked with a coalition to negotiate against the WRDPC and formed a new Council for Countermeasure for Resettlers in 1965.

After repeated requests over several years to halt construction, the village of Okawa finally agreed to an inquiry about dam construction in 1966. This agreement was reached because it had become difficult to consolidate the diverse opinions of the many people living in Okawa and other affected areas. There was added pressure placed on those living in Okawa because other affected areas, including the towns of Tosa and Motoyama, had agreed to investigation in 1964. In addition, the government of Kouchi Prefecture began to invest serious effort into methods that would secure provisions for those forced to resettle elsewhere. The prefectural government initiated promotion of a development plan in the area and began consultations with those residing in the affected area (Mizushigen Kaihatsu Koudan, 1979). However, the prefectural Assembly responded to these actions with displeasure because the prefecture did not reveal any position regarding dam construction until July 1965. The preparatory work began in 1967 and construction was completed in 1973.

Compensation schemes for the Sameura Dam resettlement

The compensation standards for the Sameura Dam construction were settled in 1967 after more than 100 instances of negotiation (Okawa Mura Shi Tsuiroku Hensan Iinkai, 1984). With regard to individual compensation, cash compensation equivalent to the material loss was paid in accordance with the standards. A group of negotiators from each affected village and town was put in place and each affected area sent an equal number of representatives to the negotiation table. One of the reasons that there was such a long negotiation period was the skyrocketing prices of land in the planned resettlement area adjacent to dam site (i.e. the towns of Tosa and Motoyama). The proposed compensation standards could not cover the price of alternative land in these neighbouring areas (Mizushigen Kaihatsu Koudan, 1979). Many people in the affected areas would not agree to the proposed compensation because they never could have found comparable land in the resettlement area for the price that they were being offered to give up their existing land. At the same time as the negotiation for individual compensation standards was taking place, the land condemnation for a new road within Motoyama town was settled. After the land condemnation was settled, this was considered a guide for compensation. As a result, the negotiation for individual compensation resumed.

The negotiation for public compensation, including public facilities and residential complexes, began in 1967; this was after the individual compensation negotiation was initiated in 1966 (Okawa Mura Shi Tsuiroku Hensan Iinkai, 1984). The letter of agreement on public compensation (including the replacement of roads and the development of a new residential area) was finally signed in 1968.

At that time the villagers requested a special subsidy for community rehabilitation that would include development of an agricultural park and a lodge. But the village of Okawa could not attain the desired subsidy to rehabilitate their livelihood because the dam construction began before the Law Concerning Special Measures in Water Resource Areas came into force in 1973. Consequently, resettled individuals only received payment for their material loss and did not receive any additional compensation to secure their future livelihood.

Lessons learned

Resettlement negotiations

The negotiation for public compensation throughout the Kusaki Dam Project was settled before the negotiation for individual compensation began. The public compensation allowed individuals much freedom to relocate near the dam site. The stakeholders collaborated to obtain comprehensive compensation deals from the WRDPC compensation schemes and there was a measure of coherence in the villagers' thinking as well as having their leader present for negotiations. The Kusaki project demonstrates that it is possible to generate acceptable compensation and therefore it is important to allow time to discuss real needs after the resettlement and to communicate these needs to the dam developers. Also, the structural negotiation units in this village successfully supported the progress of a step-by-step negotiation. Consequently, the active involvement of villagers in the consultation meeting allowed them the opportunity to discuss the future of their village.

On the other hand, negotiations concerning individual compensation in Okawa were settled by WRDPC in their haste to expedite construction. After the standard for individual compensation was announced, the villagers began to leave. A committee and the village office had then to discuss issues of public compensation for the redevelopment of Okawa Village with the dam developers. Because the plans for regional development were unclear and some villagers had already begun to leave, the remaining villagers were perplexed about whether they should also leave the village. In view of the unexpected increase in those leaving the village after the individual compensation was determined, the plan to develop the dam-affected regions had to be scaled back.

Compensation schemes

The measures taken for livelihood rehabilitation are key factors that allow resettled individuals to remain living near a dam site. Without a vision for future development, the resettled individuals find it difficult to determine whether or not they want to remain near the dam site.

Villagers affected by the Kusaki Dam Project formed the Alliance for Construction Resistance as an effort to consolidate their requests and to negotiate with the dam developers. The items discussed with the developers included rehabilitation of livelihood, construction of collective houses and bridges, development of an alternative quarry and the promotion of tourism (Seta gun Azuma Mura, 1998). The Alliance also considered measures for individual compensation, including houses for the weak. Based on the interviews, the measures put in place for the weak were considered a part of the provision for preventing depopulation of the village.

The village of Azuma requested a detailed list of the provisions offered in an attempt to secure the existing industries (e.g. forestry, agriculture and the stone industry). The developers secured the villagers' mobility by providing financial support for driving licenses as well as public transportation. Furthermore, almost half of the resettled individuals associated with the Kusaki project are engaged in the stone industry (Mizushigen Kaihatsu Koudan et al., 1974); therefore, it was important to secure stone materials for quarrying. Azuma requested a land grant from the government at the very beginning of the negotiations. Although the primary purposes of these measures (the land grant from the government, financial support for driving licenses) were to assure the quality of livelihood, the Kusaki case demonstrates that the measures were conducted in characteristics of public compensation. The case of Azuma suggested that the provisions to secure the existing industries consequently contributed to community preservation. As a result of these consolidated requests, Azuma held a strong position in the negotiations.

The village of Okawa, on the other hand, failed to seize the opportunity to develop a plan in the early stages because of the hasty negotiations initiated by the WRDPC. Before clear public compensation was firmly announced, the individual compensation was settled and residents moved out of the village. Therefore, the village was forced to scale down their original public compensation plan. For example, an inquiry concerning their intention to remain in the village, undertaken in the early stages of negotiation, showed that over 100 households wanted to remain in Okawa. Almost 70% of the households that stayed in Okawa were willing to move into collective houses in a new development area (Mizushigen Kaihatsu Koudan, 1979). However, many residents changed their minds later on. The assembly was forced to develop an alternative plan within this rapidly changing situation. The missed opportunity to develop a consolidated plan for the future of Okawa also caused emotional strain because villagers faced immediate decisions without many definite alternatives to consider.

Conclusion and discussion

This research has attempted to identify the factors related to compensation scheme and resettlement negotiation that have largely affected community rehabilitation and individual resettlement in a long-term sense.

The policy for resettlement and livelihood rehabilitation stipulated by the LSM was not applied in either the Kusaki Dam or the Sameura Dam project because the planning of these dams commenced prior to 1973. An investigation into the actual state of livelihood rehabilitation for the resettled individuals was conducted by the WRDPC and related institutions after the completion of both projects. The investigations revealed that the resettled individuals from both dam projects felt that the provisions for livelihood rehabilitation were not fully implemented. For example, the resettled individuals had been asked what kind of provisions were necessary for livelihood rehabilitation and therefore they had expected their voice to be heard and greater effort to be placed in job transition and resettlement (Mizushigen Kaihatsu Koudan et al., 1974; Nihon Dam Kyoukai, 1978). The measures for livelihood rehabilitation had to be implemented in order to mitigate the negative physical and mental impacts on the resettled individuals. The resettled individuals were looking for "visible compensations" as far as possible.

In spite of the absence of a definitive concept of livelihood rehabilitation in the Kusaki project, other special arrangements (i.e. support for quarry grants and driving licenses) greatly improved the livelihood rehabilitation of the residents. Also, the Alliance set up a minimum target for the negotiations, especially as compensation pertained to the weak. Without adequate compensation for livelihood rehabilitation, they could not decide whether they should remain near the dam site or move further away. The Kusaki project demonstrates that flexible financial and non-financial compensation measures serve not only as a catalyst for community rehabilitation but also as a bond for communities. Regarding the retention of social bonds and communities, the conventional approach to compensation, such as monetary compensation for material loss, may fail to secure the fullest extent of livelihood rehabilitation. One of the lessons learned from these case studies is that it is important to show the best available compensation measures for livelihood rehabilitation in advance and also to include a safety net for the weak to reduce the emotional strain on the resettled individuals and to contribute significantly to community preservation.

The process and length of compensation negotiations have an influence on long-term community retention. Both the Kusaki and Sameura projects demonstrate some great lessons for all stakeholders concerning dam construction and how compensation negotiations ought to proceed. The structural negotiation units associated with the Kusaki project (i.e. the Alliance and the committee) successfully consolidated their requests to the WRDPC. In contrast, the negotiation of individual compensation that took place with the Sameura Dam Project prior to determining the public compensation package undermined the community bond among villagers. The end result was that the village and villagers missed the opportunity to discuss their future due to rapid depopulation and the unexpected closure of the mine.

According to Maruyama's 1986 study on the correlation of the period of compensation negotiation and the degree of difficulty (difficulty is determined by examining the overall consequences of various conditions, such as the unique characteristic of the dam and of the area), the period of compensation negotiation would be influenced by the number of difficulties induced by these conditions in the process of compensation negotiation. These increased difficulties tended to result in a longer period of negotiation. The Kusaki project shows that the presence of a definite consensus will unite villagers and allow them to consolidate their requests so that there is more clarity in the resolution. Also, structural negotiation units with strong wills can overcome many difficulties. To avoid an unnecessarily long negotiation process, the structural negotiation units for resettled individuals should hold inclusive stakeholder dialogues, as these will greatly bolster the confidence of the stakeholders.

An uncertain redevelopment plan in terms of public compensation has a destructive impact on community rehabilitation in the long term. The Sameura project shows that a vague future plan will discourage individuals from remaining in the village. The best available future plan includes livelihood rehabilitation measures such as secure transportation, schools, job opportunities and public services that will mitigate the emotional strain on resettled individuals. Therefore, it is best to communicate the livelihood rehabilitation measures in the early stages of planning. Considering the current situation of many large infrastructure developments in developing countries, these Japanese cases demonstrate how substantial provisions for livelihood rehabilitation play a significant role in the preservation of community. Therefore, it is important to create this system in the early stages of negotiation. Moreover, it is not only important to use the conventional standards for compensation, but also best to demonstrate flexible livelihood rehabilitation measures early in the planning stage.

The Kusaki project demonstrates that, as perceptions change about dam construction projects throughout the negotiation process, several positive effects can be seen in community rehabilitation. Some of following reasons why villagers agreed to the dam construction can be seen from interviews. (1) The villagers could negotiate in their favour and take time to prepare for the resettlement through their active participation in the negotiation. (2) The majority of resettled individuals who were engaged in the stone industry were able to continue in their jobs and purchase the new heavy machinery necessary to continue their work. (3) The village of Azuma could gain a large amount of fixed property tax from the dam. Azuma and its villagers gradually saw the resettlement and dam construction as a 'new opportunity' for development in the area. The villagers believed that the public compensations would contribute to regional development and that villagers would have the opportunity to live in Azuma. In fact, Azuma has been promoting tourist businesses even since the completion of the dam. Negotiations with strong opposition parties will often lead to a deadlock; however, changes in perception can allow a reconsideration of alternative choices. Therefore, it is important to have open discussions and to learn from other dam projects throughout the process of negotiation. The affected individuals can consider their individual compensation along with the dam construction plan and decide where they will live in the short term. To garner better decisions for the resettled individuals, the affected people need to learn of the pro and con viewpoints from other experiences. The developers typically provide only a passive opportunity to learn from the experiences of others because of a risk that it may complicate or negatively influence the negotiations. Instead, the developers should support and provide opportunities to create better discussion among villagers; continuous discussions between the developers and villagers will result in better solutions and indirectly contribute to community retention.

The long-term effect of the compensation measures on community rehabilitation must be considered because the resettled individuals' surrounding environment will change in a dramatic way after the dam project. Dam construction often exposes various issues of an area such as depopulation and a weak economic basis. Depopulation of villages in mountainous or marginal areas often results in further degradation of a weak economic foundation. In these cases the possible tactics of community preservation based on the economic foundation are very limited. The Kusaki project shows that the provision of flexible compensation measures, along with a consideration of the economic foundation and geographical environment, is one of the most effective measures in preserving a community. In addition, minimal job changes also contribute to the preservation of the traditional community. A flexible approach to compensation measures may be better suited to responding to the changing socio-economic situation in the long term.

Finally, further studies need to be carried out to examine the long-term effect on resettled individuals who move to new towns and cities. Those who move to cities are often forced to convert to a new job, which requires training; these resettled individuals may have different perspectives when it comes to compensation measures compared to those who remain near their original village. The lessons learned from these Japanese cases can provide better alternative choices for resettled individuals and can also encourage governments in developing countries, international donors, and developers alike to create adequate compensation schemes when considering large-scale infrastructure development.

Note on Japanese sources

The laws in Japan cited in this article are available from the websites given in the References, but only in Japanese. The text of these laws may be found in English in the following articles.

Land Expropriation Law

Lum, M. L. L. (2007). A comparative analysis: Legal and cultural aspects of land condemnation in the practice of eminent domain in Japan and America. *Asian-Pacific Law & Policy Journal, 8*, 456–484.

Law Concerning Special Measures in Water Resources Areas

Nakamura, M. (1990). Freedom of economic activities and the right to property. *Law and Contemporary Problems, 53*(2), 1–12.

Law Concerning Special Measures on Biwako Integrated Development Project

Morino, K. (N.d.). *Wastewater management in the Lake Biwa basin*. Retrieved from http://www.nilim.go.jp/lab/bcg/siryou/tnn/tnn0264pdf/ks0264023.pdf

Multiple Purpose Land Development Law

Nakai, M. (2006). Japanese experiences on development and implementation of water resources plan at a river basin level. *Second thematic workshop on water allocation and water rights*, 5–9 June 2006, Manila, Philippines. Retrieved from http://www.narbo.jp/narbo/event/materials/twwa02/tw2_6-2.pdf

Acknowledgements

This research was conducted with the support of the Foundation of River & Watershed Environment Management. The research carried out for this study was also partially funded by the Mitsui & Co., Ltd, Environment Fund and KAKENHI (24310189). Thanks to this support we were able to collect valuable opinions and lessons from the villages of Azuma and Okawa as well as the towns of Motoyama and Tosa. We have gained great appreciation of these resettled individuals who have offered their precious time for our research. Without their valuable opinions and lessons it would have been impossible to complete this study.

References

Government of Japan (1950). *Kokudo Sougou Kaihatsu Hou* [Multiple Purpose Land Development Law]. Retrieved from http://www.bousai.go.jp/kishin/law/022.html

Government of Japan (1951a). *Tochi Shuuyou Hou* [Land Expropriation Law]. Retreived from http://law.e-gov.go.jp/htmldata/S26/S26HO219.html

Government of Japan (1951b). *Tone Tokutei Chiiki Sougou Kaihatsu Keikaku* [Comprehensive Development Plan at Tone Specific Area]. Tokyo, Japan.

Government of Japan (1960). *Shikoku-Region Development Promotion Act* [Shikoku Chihou Kaihatsu Sokushin hou]. Tokyo, Japan.

Government of Japan (1962a). *Guidelines on standards for compensation for losses associated with the acquisition of land for public purposes* [Koukyou Youchi no Shutoku ni tomonau Sonshitsu Hoshou Kijun Youkou]. Tokyo, Japan.

Government of Japan (1962b). *Memorandum on the implementation of the guidelines on standards for compensation for Losses Associated with the acquisition of land for public purposes* [Koukyou Youchi no Shutoku tomonau Sonshitsu Hoshou Kijun Youkou no Shikou ni tsuite]. Tokyo, Japan.

Government of Japan (1967). *Guidelines on the standards for public compensation associated with the implementation of public works projects* [Koukyou Jigyou no Shikou ni tomonau Koukyou Hoshou Kijun Youkou]. Tokyo, Japan.

Government of Japan (1972). *Biwako Sougou Kaihatsu Tokubetsu Sochihou* [Law Concerning Special Measures on Biwako Integrated Development Project]. Tokyo, Japan.

Government of Japan (1973). *Suigen Chiiki Tokubetsu Taisaku Sochi Hou* [Law Concerning Special Measures in Water Resources Areas]. Retrieved from http://law.e-gov.go.jp/htmldata/S48/S48HO118.html

Hattori, A., & Fujikura, R. (2009). Estimating the indirect costs of resettlement due to dam construction: A Japanese case study. *International Journal of Water Resources Development*, 25, 441–457.

Hanayama, K. (1969). *Hoshou no Riron to Genjitsu* [Theory and practice in compensation]. Tokyo: Keiso Shobo.

Maruyama, T. (1986). *Dam Hoshou to Suigen Chiiki Keikaku* [Dam compensation and headwater regional development plan]. Tokyo: Nihon Dam Kyoukai Kenkyu Bu [Research Division, Nihon Dam Association].

Mizushigen Kaihatsu Koudan [Japanese Water Agency] & Nihon System Kaihatsu Kenkyuujo [Systems Research & Development Institute of Japan] (1974). *Kusaki dam Kensetsu ni okeru Suibotsu Iten Setai no Seikatsu Saiken Jittaichousa Houkokusho* [Study report on the current status of the rehabilitation of livelihood by submerged households in the case of Kusaki Dam construction]. Tokyo, Japan.

Mizushigen Kaihatsu Koudan (1979). *Sameura Dam Kouji Shi* [Records of construction of Sameura Dam]. Tokyo, Kyodo Print.

Nihon Dam Kyoukai [Foundation for Japan Dam Association] (1978). *Sameura Dam Kensetsu niyoru Chiiki Shakai no Henka to Suibotsu Itensha he no Eikyou ni tsuite* [Influence to the community transformation and resettlers by Sameura Dam construction]. In *Dam Kensetsu to Suibotsu Hosyou*, vol. 3. [Dam constructions and compensations measures] (pp. 81–129). Tokyo: Nihon Dam Kyoukai.

Okawa Mura [Okawa Village] (1981). *Okawa Mura Shiryou, Vol. 3* [Okawa Village document]. Kouchi-ken Tosa gun Okawa mura Komatsu Okawa Kyouiku Iinkai [Educational Board of Okawa Village].

Okawa Mura Shi Tsuiroku Hensan Iinkai [History of Okawa Village Revision Committee] (1984). *Okawa Mura Sonshi Tsuiroku* [History of Okawa Village Okawa Mura [Okawa Village], Japan.

Seta gun Azuma Mura [Azuma Village, Seta County] (1998). *Seta gun Azuma Mura Shi- Tsushi- hen* [General history of Azuma Village, Seta County]. Seta Gun Azuma Mura Shi Hensan Shitsu [Editing office in Azuma Village, Seta County].

Takesada, N. (2009). Japanese experience of involuntary resettlement: Long-term consequences of resettlement for the construction of the Ikawa Dam. *International Journal of Water Resources Development*, 25, 419–430.

Index

Note: Page numbers in **bold** type refer to
figures
Page numbers in *italic* type refer to *tables*